FAITH AND FACEBOOK

EVANGELIZING A RURAL PARISH COMMUNITY IN SOUTHEAST MISSOURI THROUGH SOCIAL MEDIA

Rev. Dominic Ibok, D.Min.

En Route Books and Media, LLC
Saint Louis, MO

ENROUTE
Make the time

En Route Books and Media, LLC
5705 Rhodes Avenue
St. Louis, MO 63109

Cover credit: Sebastian Mahfood

ISBN-13: 978-1-956715-33-0
Library of Congress Control Number: 2022934046

CONTENTS

FIGURES

ACKNOWLEDGMENTS

I want to acknowledge and thank all members of Sacred Heart Parish, Caruthersville, and Immaculate Conception parish, New Madrid, for their support throughout the planning and writing of this project. I am indebted to my Bishop, Edward Rice, and members of the Catholic Center, Sarah and Jan. They have supported me throughout this academic journey at the Aquinas Institute of Theology.

I also express my gratitude to my thesis advisor Sr. Sara Fairbanks, O.P., and my reader, Dr. Sebastian Mahfood, O.P., for all the time and energy they devoted to helping me improve this thesis. Mainly, my thanks to Sr. Sara for her patience and direction throughout this process. I thank my editors, Rev. Donald Goergen, O.P., Teresa, Kate, Emily, and Ms. Loretto, for their work on my thesis and helping me with my writing. They helped me develop a love for writing.

I thank the Dominican Province of St. Albert the Great for allowing me to study and grow in the art of preaching. Finally, I thank the Lord for his grace and the outpouring of his gift of Wisdom that continues to sustain me in my preaching. To the Lord be all glory and praise, and to Our Blessed Mother be all honor, forever, Amen.

ABSTRACT

FAITH AND FACEBOOK: EVANGELIZING A RURAL PARISH COMMUNITY IN SOUTHEAST MISSOURI THROUGH SOCIAL MEDIA

Ibok, Dominic O., M.A., M.Div., D.Min. Aquinas Institute of Theology, Saint Louis, Missouri, 2021.

In Southeast Missouri, where Catholics are a minority, the challenge of diminishing participation and reduced membership requires an intentional engagement of the inactive members and the unaffiliated within the community. Unfortunately, many members of the laity do not have the awareness or skills to evangelize. In a rural community, the challenges seem more significant due to limited personnel and resources.

This thesis project aims to equip the laity with the skills and tools to evangelize inactive Catholics and the unaffiliated. It will involve growing in prayer, study, generosity, evangelization, and the discernment of their charisms. With the benefit of social media engagement like Facebook live stream, an opportunity to engage and evangelize is available for our rural parish community. The resources from experienced lay evangelists in evangelization and social media engagement will provide a template that can enhance the development of a program on evangelization for my rural parish community.

To test the effectiveness of these resources, seventeen participants engaged in a nine-week program to learn how prayer, study, generosity, evangelization, the discernment of their charisms, and teamwork can prepare them for evangelization. This thesis engaged the inactive Catholics and the unaffiliated in rural Southeast Missouri by using Facebook live stream as an evangelization tool. The focus of this project is thus reminding active Catholics of their responsibility to evangelize and how utilizing a familiar social media portal like Facebook can enhance the process even in a rural community.

Chapter 1
The Backstory

1.1: Introduction

As a seminarian in formation about twenty years ago, I had the opportunity of engaging in door-to-door evangelization in the inner city of St. Louis. The experience challenged my perceptions about lay evangelization and engagement since it was one of a few opportunities I had working alongside active Catholics who were intentional in sharing their faith. In my time as pastor in Southeast Missouri, I see similar opportunities for evangelization and engagement. My observation is that there is great potential in the laity in this region to engage both inactive Catholics and the unaffiliated to appreciate the Catholic Faith. My pastoral experience in this region continues to be a platform for engagement and creates the backstory for this thesis project on lay evangelization in the Bootheel. Moreover, southern Missouri is part of the "Bible belt" because of the Southern Baptist's heavy presence and the Assemblies of God denominations—Protestant denominations that emphasize a fundamentalist approach to the Scriptures. Furthermore, the Catholic presence here though minor, still provides Catholics within this region an opportunity to grow in their faith and engage their protestant and unchurched friends within their communities.

Southeast Missouri falls under the diocese of Springfield-Cape Girardeau, which was established in 1956. However, in the diocese's geographical area, which includes thirty-nine counties in the southern part of the State, 66 parishes with approximately 60,000 Catholics make up about 2 percent of the population. This number is much smaller when you come to the bootheel, located in the southeast of the diocese. In this region, the Catholic population is about 0.96% or about 600 Catholics.

I am currently Pastor of two rural parishes in this region of the diocese of Springfield-Cape Girardeau, Missouri. The parishes are Immaculate Conception Parish, New Madrid, and Sacred Heart Parish, Caruthersville. My thesis project will focus on Sacred Heart parish located in Pemiscot County in Southeast Missouri. The parish presently has 91 registered families that consist of approximately 144 individuals. The city is one of the poorest in the State. The challenge this presents for a rural parish community is the ongoing migration or departure of young families and persons searching for better economic and financial opportunities. Since the 2006 tornado that damaged or destroyed about 60 percent of the town, many have moved to nearby cities, searching for better living situations. In the past ten years, beginning November 2010 till November 2020, 45 families have moved out of the parish, and about 39 members have become inactive within the same period. The Catholic Faith is not experiencing any significant growth within the parish boundaries.

Therefore, the context is such that many active members continue to express concern about the parish's future regarding its membership due to families' ongoing migration and diminishing participation. Moreover, while members express these concerns, there is a lack of awareness of how to engage or evangelize inactive members to increase their participation or outreach to the unchurched within the rural community. From my observation, the lack of engagement and evangelization in the parish will only decrease participation and membership.

My work as a pastor often involves a variety of ministerial tasks that include evangelization. The work of evangelization is a vitally important ministry and comes as a direct command from Jesus, who reminds his first disciples to go out into the world and proclaim the Good News. To be done effectively, however, we need parishioners with various gifts and talents to be actively involved. As *Lumen gentium* states, "Pastors also know that they themselves are not meant by Christ to shoulder alone the entire saving mission of the Church toward the world."[1]

---

[1] Vatican Council II, Dogmatic Constitution on the Church *Lumen gentium* (21 November 1964) 30, in *Vatican Council II Constitutions, Decrees and Declarations: A Completely Revised*

Our parish community in Pemiscot County has seen its membership dwindle over the years. Nevertheless, a few active members continue to show a high degree of commitment to help find ways to evangelize by acting as collaborators but express a desire to learn how to engage inactive members and the unaffiliated. As a parish community, to respond to this problem, members need to be aware that "there is no true evangelization if the name, the teaching, the life, the promises, the Kingdom and the mystery of Jesus of Nazareth, the Son of God are not proclaimed. The history of the Church, from the discourse of Peter on the morning of Pentecost onwards, has been intermingled and identified with the history of this proclamation."[2] Simultaneously, the concept of evangelization may seem commonplace. Not every Catholic who comes to Church every weekend recognizes the need to be part of this mission, and not many will volunteer to engage others in the faith.

Within the parish community, parishioners need tools and skills to engage inactive members and the unchurched. Nevertheless, depending on each person's faith formation, the willingness to participate in any effort may be strong or

---

Translation in Inclusive Language, edited by Austin Flannery (Northport NY: Costello Publishing Co., 1996).

[2] Paul VI, Apostolic Exhortation *Evangelii nuntiandi* (8 December, 1975), (Frederick, MD: The Word Among us Press, 1975), 24.

weak. The parish community of Sacred Heart in Caruthersville recognizes that while prayer is an essential aspect of our mission, there is also the need to be intentional in our efforts to reach out to inactive members and the unchurched. In doing this, an awareness that "evangelization happens when the word of Jesus speaks to people's hearts and minds. Needing no trickery or manipulation, evangelization can happen only when people accept the Gospel freely, as the 'good news' it is meant to be, because of the power of the gospel message and the accompanying grace."[3] In engaging others through evangelization, our rural parish community lives out her mission of sharing the gospel message and making Jesus Christ present to members who have become inactive or persons within the rural community who may be seeking to encounter Jesus. In a rural parish community, as in an urban parish community, there is a constant need to remind members of the necessity of evangelization.

---

[3] United States Conference of Catholic Bishops, *Go and Make Disciples: A National Plan and Strategy for Catholic Evangelization in the United States* (Washington D.C.: The Printery House, 2002), 11.

1.2: Problem

The problem that my thesis will address is that the parishioners are unaware of their call to evangelize. Simultaneously, some have reservations about the parish's future due to diminishing attendance and not knowing how to address the issue. These reservations about the parish and possible steps for the future have led to discussions about how to evangelize and establish a process to engage inactive members and the unchurched in the faith to build up parish membership and participation. The discussion about increasing membership and participation is critical for the life and growth of the parish community. The United States Conference of Catholic Bishops challenges the clergy to invite the laity to reflect and act accordingly in these words:

> Today Christian Churches must communicate the importance of the laity's witness and service within the family and within the professional, social, political, and cultural life of society. An effective parish or congregation will help its members make connections between worship and work, between liturgy and life in the family, community, and workplace. For this reason, church ministers—are called to strengthen and equip lay people to be wit-

> nesses to Christ, acting in the power of him who is the Good Shepherd and humble servant of all.[4]

The laity within our rural parish community needs some orientation and education of their vocation as disciples and some guidance on evangelization. However, within our rural parish community, making the connection to increase participation and attract membership will be essential to developing this thesis project. Moreover, to arrive at the point of witnessing to others and heeding the call of the bishops' conference in our rural community, there is the need to equip the laity and remind them of their responsibility in the mission of the Church. Recognizing the importance of evangelization within a rural parish community can be an uphill challenge due to the various assumptions Catholics living in a fundamentalist context may experience frequently. Some of these assumptions arise from a fear of sharing one's faith; another contributing factor could be a lack of exposure to evangelization benefits regarding a rural parish community and how this can be transformative and enriching.

Engaging the problem of how and why evangelization is necessary requires the collaboration between clergy and

---

[4] United States Conference of Catholic Bishops, *Called and Gifted for the Third Millennium* (Washington D.C.: USCC Inc., 2001), 16.

laity to develop a strategy that will engage active Catholics, inactive Catholics, and the unaffiliated. In its message on World missions, the National Conference of Catholic Bishops in the United States notes, "The Church extends an invitation, realizing that others may not respond. If we extend our invitation well, witnessing in the love of God and the image of Christ and the fire of the Holy Spirit, we have fulfilled Christ's mandate. Acceptance of the Gospel and conversion to Christ is the working of God's grace, a mystery beyond comprehension which we accept in faith."[5] A notable aspect of the bishops' statement is evident in our rural parish community is the absence of an invitation to non-Catholics or an intentional engagement of inactive members. When we fail to invite others to an encounter of faith with Jesus Christ in the Church, growth and community participation will see a drop off over time. In devising a means to engage the problem of this thesis project, we need to be mindful that "the work of evangelization flows from who we are in Jesus Christ—from our baptismal vocation and ongoing encounter with the Lord."[6] If the laity receives instruction on their baptismal vocation, there is a chance

---

[5] National Conference of Catholic Bishops, *To the Ends of the Earth: A Pastoral Statement on World Mission*(Wshington D.C.: Office of Publishing and Promotion, 1986), 14.

[6] Curtis Martin, *Making Missionary Disciples: How to live the Method Modeled by the Master* (Genesee, CO: FOCUS, 2018), 9.

that there will be a desire to evangelize and draw others into an encounter with Jesus Christ.

It is important to note that a desire to evangelize within our rural parish community reflects a need for renewal within our parish community. In accord with this desire for renewal, Pope Francis notes:

> Today, as the Church seeks to experience a profound missionary renewal, there is a kind of preaching which falls to each of us as a daily responsibility. It has to do with bringing the Gospel to the people we meet, whether they be our neighbors or complete strangers. This is the informal preaching which takes place in the middle of a conversation, something along the lines of what a missionary does when visiting a home. Being a disciple means being constantly ready to bring the love of Jesus to others, and this can happen unexpectedly and in any place: on the street, in a city square, during work, on a journey.[7]

Evangelization, accompanied by an intentional desire to renew our parish, is new for our rural parish community in

---

[7] Francis, Apostolic Exhortation *Evangelii gaudium* (24 November 2013), (Washington D.C.: United States Conference of Catholic Bishops, 2013) 127.

Southeast Missouri. Reaching out to inactive members amidst their challenges is essential as a starting point. Trying to reassure and invite them to return home to active participation may be challenging but also necessary.

Moreover, evangelization is not limited to those outside the church but requires engaging members who have become inactive within the parish community. In other words, evangelization is for all peoples as St. John Paul II notes that

> The Church was "sent by Christ to reveal and communicate the love of God to all people and nations." This mission is one and undivided, having one origin and one final purpose, but within it, there are different tasks and kinds of activity. First, there is the missionary activity which we call *mission ad gentes*, in reference to the opening words of The Council's decree on this subject. This is one of the Church's fundamental activities: it is essential and never-ending. The Church, in fact, "cannot withdraw from her *permanent mission of bringing the Gospel* to the multitudes and millions of men and women who as yet do not know Christ the Redeemer of humanity. In a specific way, this is the

missionary work which Jesus entrusted and still entrusts each day to his Church."[8]

The mission to all people, however, begins with our primary mission to our local parish community. Renewal for our parish community involves instruction and preparation of all the members for the work of evangelization.

1.3: Purpose

My project aims to address the lack of evangelization and provide training and resources to increase participation and membership within the parish. The process will involve developing a training program that will raise the awareness of evangelization and enhance the development of intentional discipleship skills within and outside the parish community. Through my studies, I will develop a program that will enhance members' desire to evangelize. Through faith formation, members will learn skills that promote teamwork for evangelization, collaboration, ways in which they can develop in their prayer lives, and how they can become intentional in the practice of their faith within and outside the parish community. The formation sessions will

---

[8] John Paul II, Encyclical Letter *Redemptoris missio* (7 December 1990), (Washington D.C.: United States Conference of Catholic Bishops, 1990) 31.

include drawing resources from internationally known evangelists and founders. Matthew Kelly is the founder of the Dynamic Catholic Institute, and Sherry Weddell is the co-founder of the Catherine of Siena Institute. The Siena institute engages the laity in discerning their gifts to prepare them for evangelization and mission in the Church. Their work will provide our rural community with some practical tools to enhance our evangelization efforts.

1.4: Formation and Catechesis

Evangelization in every context will require a different approach; however, the expectation is that as disciples, we need to be grounded in our faith. One of the ways we prepare ourselves is through catechesis and faith formation programs. These programs can involve reviewing the basics of our faith in the catechism of the Catholic Church or engaging contemporary resources available today. Incorporating these faith formation programs can motivate active Catholics who desire to evangelize and engage inactive Catholics or the unchurched. Formation and catechesis for evangelization strengthen each member with the knowledge necessary to evangelize. Pope Paul VI, on engaging Catholics, notes:

> The Church is an evangelizer, but she begins by being evangelized herself. She is the community of

> believers, the community of hope lived and communicated, the community of brotherly love, and she needs to listen unceasingly to what she must believe, to her reasons for hoping, to the new commandment of love. She is the People of God immersed in the world and often tempted by idols, and she always needs to hear the proclamation of the "mighty works of God," which converted her to the Lord; she always needs to be called together afresh by Him and reunited. In brief, this means that she has a constant need of being evangelized if she wishes to retain freshness, vigor and strength in order to proclaim the Gospel.[9]

Our formation's strength is foundational for our work as witnesses in evangelizing and inviting inactive members to active participation in the Church's life. In other words, this formative step will provide for all the faithful a twofold goal of renewal and preparation. In engaging in such a formative process that renews and prepares, members become more established as witnesses. To this point, Pope John Paul II states:

> People today put more trust in witnesses than in teachers, in experience than in teaching, and in life

---

[9] EN 15.

> and action than in theories. The witness of a Christian life is the first and irreplaceable form of mission: Christ, whose mission we continue, is the "witness" *par excellence* (Rv 1:5; 3:14) and the model of all Christian witness. The Holy Spirit accompanies the Church along her way and associates her with the witness he gives to Christ (cf. Jn 15:26-27). The first form of witness is *the very life of the missionary, of the Christian family*, and *of the ecclesial community*, which reveals a new way of living. The missionary who, despite all his or her human limitations and defects, lives a simple life, taking Christ as the model, is a sign of God and of transcendent realities. But everyone in the Church, striving to imitate the Divine Master, can and must bear this kind of witness; in many cases, it is the only possible way of being a missionary.[10]

In expressing the importance of our call to evangelize, our Holy Father stresses that as witnesses, we are also missionaries whose proclamation strengthens the parish community's life. Our missionary call aligns with our baptism as Pope Francis writes:

---

[10] RM 42.

> In virtue of their baptism, all the members of the People of God have become missionary disciples (cf. *Mt* 28:19). All the baptized, whatever their position in the Church or their level of instruction in the faith, are agents of evangelization, and it would be insufficient to envisage a plan of evangelization to be carried out by professionals while the rest of the faithful would simply be passive recipients.[11]

Discipleship challenges us to evangelize outside the boundaries of the Church. It reminds us of our duty to seek out those who have stepped away from active participation in the Church. The Samaritan woman is an excellent example of one who became inactive. The scriptures indicate that her encounter with Jesus transformed something inside of her. According to John, she is motivated to invite others within her community to a similar encounter, as recorded in the fourth chapter of the Gospel. It is also noteworthy that for discipleship to overflow into evangelization, a daily and ongoing encounter through study and prayer is necessary for all Catholics.

---

[11] EG 120.

1.5: Discernment and Dialogue

Discerning the need for evangelization and having a concrete discussion on how to execute a pastoral plan is never an easy task. With many ideas and suggestions, there is always a need for careful planning that will include initiatives relative to our setting. In my ministry in Southeast Missouri, this involves recognizing an aging community and the importance of social media in our work. In thinking of a plan for evangelization that will challenge my rural parish community to be intentional in their approach, the instruction from Pope Paul VI seems timely when he states:

> For the Church, evangelizing means bringing the Good News into all the strata of humanity, and through its influence transforming humanity from within and making it new: "Now I am making the whole of creation new. But there is no new humanity if there are not first of all new persons renewed by Baptism and by lives lived according to the Gospel. The purpose of evangelization is therefore precisely this interior change, and if it had to be expressed in one sentence the best way of stating it would be to say that the Church evangelizes when she seeks to convert, solely through the divine power of the message she proclaims, both the personal and collective consciences of people, the activities in which they

engage, and the lives and concrete milieu which are theirs.[12]

The work of evangelization requires recognizing that proclaiming or sharing our faith is not limited to Church attendance but involves all the informal moments of our daily lives. The goal, however, remains the same; a return to the sacramental life of the Church, active participation by inactive members, and an engagement of the unchurched.

Furthermore, without engaging intentionally, the goal of increasing participation and membership will be challenging to attain. Whether it is inactive Catholics or the unchurched, "the unique claim of our message does not negate the sincerity and faith of others; likewise, the sincerity and faith of others do not take away from the clarity and truth of our message."[13] We should not hesitate to share our faith's beauty and invite the unchurched to encounter the faith we proclaim and live as Catholics.

1.6: Invitation

Some who have become inactive members often suggest that they do not feel welcome when and if they choose to return. Indeed, this is due to the feeling of being judged by

---

[12]EN 18.

[13] USCCB, *Go and Make Disciples*, 19.

some within the community. My observation is that not all Catholic communities present this challenge. Nevertheless, there is also the suggestion that some never receive an invitation to come back. Part of our strategy will be to invite all who have become inactive members irrespective of their situations. The life of Jesus demonstrates that he was willing to invite all, even those not recognized or respected by society. Yet, as Catholics, we should not assume that since one has become inactive, they still possess an automatic invitation. There still needs to be an invitation to inactive members. Every invitation creates a new opportunity for engagement. The faith of each member grows differently, and for some, the simple act of invitation can make a difference. In looking at the life of Jesus, Pope John Paul II notes that "Jesus gradually reveals the characteristics and demands of the kingdom through his words, his actions, and his own person. The kingdom of God is meant for all mankind, n d all people are called to become members of it."[14] Therefore, our hospitality or charity should begin with members who do not participate regularly and need to be re-invited back to the parish community. This act of extending an invitation to non-practicing members may have various responses but is necessary as a part of our evangelizing efforts. If we recognize that the kingdom starts from within and flows without, we should be willing to take the

---

[14] EN 14.

risk involved and offer to all our invitation to worship and active participation.

Since the kingdom is the concern of everyone, building that kingdom to which the Church is a reflection should also engage everyone—both active and inactive alike. For this engagement to be possible, evangelization must include inviting all, and in our work, this includes members who have become inactive.

1.7: Ongoing Conversations

As a parish community, an initial study of those within the parish boundaries and a discussion to determine their openness is essential. After that, an invitation through social media and communication gadgets follows where and when necessary. Our next step in evangelizing and engaging the inactive members and the unchurched is through personal conversation. Conversations can take time due to schedules and work. Still, as evangelizers, the patience of the farmer who plants a seed in anticipation of the harvest is necessary throughout this process. Ratzinger notes that "[evangelization] cannot mean immediately attracting the large masses that have distanced themselves from the Church by using new and more refined methods, it means to dare, once again and with the humility of the small grain, to leave up to God the when and how it will grow (Mk 4:26-

29)."[15] Therefore, conversations as part of evangelization remind us that we need to be willing to engage throughout the process the inactive members and non-Catholics and try and follow-up for a response.

Besides the personal conversations made as part of this process, we need to complement these conversations with some engagement through social media outreach. We try not to limit our outreach and conversations to a specific place or time but work to proclaim whenever the opportunity is present, as Pope Paul VI notes:

> Our century is characterized by the mass media or means of social communication, and the first proclamation, catechesis, or the further deepening of faith cannot do without these means, as we have already emphasized. When they are put at the service of the Gospel, they are capable of increasing almost indefinitely the area in which the Word of God is heard; they enable the Good News to reach millions of people.[16]

---

[15] Joseph Ratzinger, *The New Evangelization: Building the Civilization of Love*, accessed online December 1, 2019 at https://www.piercedhearts.org/benedict_xvi/Cardinal%20Ratzinger/new_evangelization.htm, 1.

[16] EN 45.

Therefore, in highlighting the importance of proclamation, Pope Paul VI reminds us that conversations, which enhance our proclamation, should not be overlooked or limited in their scope. In a small rural community, talks and outreach seem to be appreciated by many inactive members and the unchurched.

A simple act of invitation followed by a conversation could impact a person's desire to return to the Church or explore the Catholic faith. In thinking of conversations, the laity has an advantage because of their context and awareness of members who have become inactive or non-Catholics who may desire to engage the Catholic Faith. Working with the laity as collaborators in evangelization can attract inactive members to renew their participation or the unchurched as St. John Paul II notes that:

> Within the Church, there are various types of services, functions, ministries, and ways of promoting the Christian life. I call to mind, as a new development occurring in many churches in recent times, the rapid growth of "ecclesial movements" filled with missionary dynamism. When these movements humbly seek to become part of the life of local churches and are welcomed by bishops and priests within diocesan and parish structures, they

> represent a true gift of God both for new evangelization and for missionary activity properly so-called.[17]

The participation of the laity in any parish evangelization effort is necessary if we will make any substantial progress within the parish. Missionary activity and evangelization that includes their involvement has far-reaching effects, as indicated by our Holy Father. While the priest cannot reach many who may be desirous of returning to the faith or engaging the Catholic Faith, the presence of the laity is necessary to meet this need. As part of this process, the laity opens up a channel of communication that complements the work of the priest.

1.8: Hypothesis

My hypothesis is that the role of the laity in evangelization is indispensable and necessary for the growth and life of the Church, as the Vatican II documents indicate. I will explore how the laity responds to evangelization by preparing them to engage inactive members and non-Catholics within the community. Using Matthew Kelly and Sherry Weddell's writings and resources, this thesis project will help participants develop as a team through gospel-based faith-sharing, prayer, forming a habit of study, practicing

[17] RM 72.

generosity, and engaging in evangelization. At the end of this process, I will develop an evangelization intervention involving social media, specifically Facebook, in collaboration with the participants.

The usefulness of any pastoral plan of action is that it facilitates evangelization within the parish community. Moreover, each baptized person encounters an opportunity for faith renewal through the work of evangelization. The Church documents encourage all the faithful in evangelization and the necessity of sharing their Faith. In the document on evangelization, the Synod of Bishops expresses the following when it notes that, "Evangelization, in general, is the everyday work of the Church. With the assistance of the Holy Spirit, this so-called ordinary evangelizing activity can be endowed with renewed vigor. New methods and new forms of expression are needed to convey to the people of today the perennial truth of Jesus Christ, forever new and the source of all newness."[18] In renewing our rural parish community, evangelization helps us prepare members who will serve the body of Christ in new ways. Renewal and ongoing conversion are opportunities for strengthening the

---

[18] Synod of Bishops, *The New Evangelization for the Transmission of the Christian Faith,* accessed online December 1, 2019 at http://www.vatican.va/roman_curia/synod/documents/rc_synod_doc_20120619_instrumentum- xiii_en.html

faith of active members who will engage inactive members who choose to return to active participation.

The approach expressed in this thesis project will require several weeks of formation and the discernment of gifts by each participant. Increasing participation and membership through evangelization and collaboration with the laity is necessary for growth and renewal. The weekly participation at our weekend Masses and fellowship events will be an opportunity to gauge how members respond to our invitation. Indeed, our online presence will enhance our reaching out to the larger community outside our rural parish boundaries. The effectiveness of evangelization in our parish community will have the ability to impact our diocesan deanery and the entire diocese of Springfield-Cape Girardeau. It may seem a minor contribution to the work of evangelization, but it can also serve as a template for other rural parish communities. As the United States bishops Conference notes, "to create a culture of witness, we must live explicit lives of discipleship. Being a disciple is a challenge. Fortunately, one does not become a disciple of Christ on his or her own initiative. The work of the Holy Spirit within the Christian community forms the person as a disciple of Christ. One seeking to learn how to be a disciple of Christ does so through apprenticeship."[19] Therefore,

---

[19] Committee on Evangelization and Catechesis, United States Conference of Catholic Bishops, *Disciples Called to Wit-*

this thesis project will involve the cooperation of the entire parish to increase participation and membership of both inactive and the unchurched.

1.9: Road Map

In this project, I will explore how evangelization in rural southeast Missouri presents some significant mission opportunities. Through my ministerial intervention, I will show how evangelization can be approached or sustained, especially in a parish setting where many members have become inactive in the practice of their faith. In doing this, reference will be made to Church documents and the guidance they offer for evangelization and ministerial initiatives. After three years of rural parish ministry, I have observed how ongoing faith formation and evangelization continue to transform my parish communities. While this process has brought new evangelizing opportunities, it also presents challenges that require some creative action in ministry.

As a guide to the reader, my thesis will begin in the next chapter with a brief biblical foundation for evangelization and mission. The biblical foundation will facilitate an understanding of the mission mandate spoken by Jesus to his disciples. This chapter will explore the theological basis for

---

ness the New Evangelization (Washington D.C.: United States Conference of Catholic Bishops, 2013), 14.

evangelization and how the laity contributes to the parish's mission through evangelization. Church teachings from Vatican II documents, papal exhortations, encyclicals on the laity and evangelization, and recent documents from the United States Catholic Bishops will be a significant part of this chapter.

Chapter three of this thesis will give an overview of the key players—Matthew Kelly and Sherry Weddell—whose resources guide the development of my thesis project. I will further discuss why their contributions to my thesis project are significant and why their specific choice stands out, considering the many lay evangelists who promote the participation of the laity in evangelization.

In chapter four of my thesis, I will begin with the Church's Decree on social media—*Inter mirifica*—and highlight the interdisciplinary framework for this thesis project. In a rural parish context, social media is one creative way to develop a plan for evangelizing members and engaging non-Catholics. Pertinent to this chapter, I will discuss Meredith Gould, whose writings and experience contribute to my Social Media engagement in evangelization.

The fifth chapter of my thesis will review the nine-week process and the various interventions that were part of the thesis project. The number of participants and the project's effect on their faith in preparation for evangelization will highlight this chapter.

The final chapter of my thesis will review the data and highlight the project's impact on the rural parish community. Discussion on the effectiveness of social media in evangelization and the possibility of expanding this project to other rural parish communities within this region will highlight this chapter.

In conclusion, since evangelization entails communicating our faith in Jesus Christ, there will be a need to instruct the laity on some available means to deepen and nourish their faith. Since "language itself, the words by which we make sense of our lives and the world around us, comes to us from others, preserved in the living memory of others. Self-knowledge is only possible when we share in a greater memory. The same thing holds true for faith, which brings human understanding to its fullness."[20] The language of faith is what we need as we continue to be formed, discern, invite, engage and anticipate active participation and ongoing conversation with inactive members and the unchurched. Evangelization continues the work of propagating our faith in Jesus Christ. For a rural parish community in Southeast Missouri, recognizing our call to discipleship helps us witness our faith and is an essential first step.

---

[20] Pope Francis, Encyclical Letter *Lumen fidei* (29 June 2013), (Washington D.C.: United States Conference of Catholic Bishops, 2013) 38.

Chapter 2

The Entire People of God called to Evangelize

2.1: Introduction

In the first chapter, I briefly described my ministerial context in rural Missouri. Also, I noted that my thesis would address the problem that the parishioners are unaware of their call to evangelize. Simultaneously, some members have reservations about the parish's future due to diminishing attendance and not knowing how to address the issue. This project aims to address the lack of evangelization and provide training and resources to increase participation and membership within the parish. As stated in chapter one, the process will involve developing a training program that will raise the awareness of evangelization and enhance intentional discipleship skills within and outside the parish community. In chapter one, my hypothesis is that the role of the laity in evangelization is indispensable for the growth and life of the Church, as the Vatican II Documents indicate. I will explore how the laity responds to evangelization by preparing them to engage inactive members and the unaffiliated within the community.

In this chapter, I will provide the biblical and theological foundation for the project, giving particular attention to the importance of the divine mandate to go and evangelize. In the first section, I will highlight the scriptural basis for the Church's call to evangelize under the power of the Holy Spirit, who gifts and guides the faithful in their mission to proclaim the good news.

In the second section, I will discuss the documents of Vatican II and how the laity's participation in evangelization is essential for parish renewal and the Church's mission. I will draw from *Lumen gentium* to highlight The Council's innovative teaching on the identity and evangelical mission of the laity. I will address some concepts from *Dei Verbum, Gaudium et spes, Apostolicam actuositatem,* and *Ad gentes* that strengthen and support the laity's involvement in the Church's evangelization mission.

In the third section, I will discuss some contemporary papal encyclicals on evangelization, namely, *Evangelii nuntiandi, Redemptoris missio, and Evangelii gaudium,* as well as some directives from the United States Conference of Catholic Bishops, including *Co-Workers in the Vineyard of the Lord* and *Go and Make Disciples.* Lastly, I will address the implications these writings have for my project regarding the promotion of lay evangelization in my rural parish community.

2.2: Divine Mandate to Evangelize

In a close reading of each of the four Gospels, a common occurrence is that Jesus Christ calls disciples to be his companions as he embarks on his public ministry. As the Gospels unfold, these disciples go through a series of informal formations and later become the instruments or channels of the Good News. In his book, *The Church According to the New Testament*, Daniel Harrington notes:

> The gospel passages about Jesus' first disciples remain rich resources for the Church today. The positive and enthusiastic response to Jesus by his first followers remains paradigmatic for all Christians. The idea of discipleship as being with Jesus and sharing in his ministry of proclaiming the kingdom of God and bringing his healing power to those in need provides the best definition of the vocation of the Christian today. At the same time, the Evangelists' frank admission of the disciples' failures is a sobering challenge to face up to the difficulties involved in following the way of Jesus of Nazareth.[21]

---

[21] Daniel Harrington, S.J., *The Church According to the New Testament: What the Wisdom and Witness of Early Christianity Teach Us Today* (Franklin, WI: Sheed and Ward, 2001), 11.

In his assessment of the first disciples, Harrington is quick to note that the encounter of the first disciples with Jesus provides a teaching moment on discipleship and how that affects our mission to share the good news. Harrington further notes that:

> The task of the Church in the world is to believe, proclaim, and live out the message of Jesus about the kingdom of God. By itself and through its own efforts, the Church does not and cannot bring about the fullness of God's kingdom. In fact, the New Testament writers avoid statements about our building God's kingdom or bringing it about—a silence that reflects sound Jewish and early Christian theology that should be respected. It is not our kingdom to build or to bring about. Moreover, the Church should not make itself the focus of its own preaching. Rather, the New Testament writers make God's kingdom and Jesus central.[22]

The kingdom of God and our encounter with Jesus Christ are essential for our vocation as disciples. The graces we receive in baptism remind us of our ultimate end, which is the kingdom of God. A closer look at what it means to be a disciple is essential for considering our mission to evange-

---

[22] Ibid., 20.

lize, as is evident in Anthony Gittins book; *The Way of Discipleship*, where he writes:

> Discipleship is not something people can simply fashion for themselves, and so there can be no self-taught disciples. Jesus is perfectly clear about the nature and purpose of discipleship: he himself is the initiator, and he looks to potential disciples for their response to his call and initiative. Disciple is not a self-defining word and does not identify a totally self-sufficient person, for no disciple is an independent agent. Disciples are interdependent, first in relation to Jesus and then in relation to each other. They form a community, united in faith and mission, if not in a specific locality.[23]

Since the disciple's vocation does not make the disciple independent but requires an ongoing relationship with Jesus Christ, it is essential to strive to imitate the example of the first disciples—the twelve and the seventy—who provide a pattern for discipleship. Gittins adds:

---

[23] Anthony Gittins, *The Way of Discipleship: Women, Men,and Today's Call to Mission* (Collegeville, MN: Liturgical Press, 2016), 41.

> We should note from the outset that in the New Testament, the word" disciple"—as indeed the word "apostle"—is used (confusingly) both in a narrow, restricted sense and also in a broader, more inclusive sense. In the former sense, it applies specifically to those who are called by Jesus, develop a personal relationship with him and, form a community devoted to proclamation of the realm of God. In Luke's gospel, particularly, the twelve disciples are routinely called apostles. However, in the broader sense, "disciple" applies to anyone who shows real faith and trust in Jesus and whose encounter with him produces a conversion to the Way of Jesus. This is why we ourselves can claim to be latter-day disciples of Jesus.[24]

Therefore, Gittins's writing clarifies the use of the word "disciple" as it applies to all Christians. Moreover, Gittins further illustrates that discipleship within the New Testament has some standard features that apply to discipleship's vocation for all times. He indicates:

> Being on the lookout for the way different people address Jesus or indeed approach him wordlessly, we may more readily appreciate the dynamics of

[24] Ibid., 47.

> discipleship. But there is at least one more thing we can identify, and that is what we might call the stages of discipleship. Again, it is more a matter of looking across the whole panorama of the New Testament, specifically the gospel stories, to see if we can perceive or deduce a pattern. Discipleship does indeed have a shape or form; it is not something vague or evanescent, and when Jesus looks for potential disciples, he does not do so in some random fashion. He is looking for certain qualities and a certain disposition that, frankly, not everyone possesses. But he is not looking for clones, and there is no such thing as a "one-size-fits-all" discipleship. As we look across the full range of gospel stories, it seems that mature discipleship is the result or outcome of a relationship between Jesus and another person that develops in three stages. First, there is call or an encounter; then, there is a disturbance or a displacement; and, finally, there is a sending forth or a "co-missioning."[25]

Indeed, Jesus is the source and center of our vocation as disciples, and the call to discipleship has its responsibilities. In looking at the New Testament writings on discipleship, the final chapter of the Gospel according to Matthew gives

[25] Ibid., 58.

some indication of how Jesus instructs his disciples to continue the task of making disciples of every nation.

The Gospel according to Matthew ends by highlighting how Jesus commissions his disciples and instructs them to go out and spread the good news. The passage from the *New American Bible* is as follows:

> The eleven disciples went to Galilee, to the mountain to which Jesus had ordered them. When they saw him, they worshiped, but they doubted. Then Jesus approached and said to them, "All power in heaven and on earth has been given to me. Go, therefore, and make disciples of all nations, baptizing them in the name of the Father, and of the Son, and of the Holy Spirit, teaching them to observe all that I have commanded you. And behold, I am with you always, until the end of the age."[26] (Matt. 28:16-20)

These words of Jesus at the end of the Gospel are foundational when thinking of discipleship and evangelization. The text setting takes us to a familiar city where he began his public ministry—Galilee. The disciples gather in response to the instruction by Jesus to the women who first met the risen Jesus at the empty tomb. There Jesus empow-

[26] NAB, Matt. 28:16-20.

ers them to continue the work he initiated through his words, action, and life.

In his book: *The Gospel of Matthew*, Donald Senior notes that "Scholars have debated what literary form this passage may be patterned after, describing it for example as a 'theophany' or as similar to an Old Testament commissioning text."[27] According to Matthew, the disciples' mission was only to the Jews at the start of Jesus' ministry in the Gospel. However, by the end of Jesus' earthly life, he extends it to all nations in the known world. In his book, *The Churches the Apostles Left Behind,* Raymond Brown explains that, unlike Luke, who wrote a second book to describe the developments in the post-resurrectional Church, Matthew interweaves what happened in the post-resurrectional Church into the account of Jesus' public ministry.[28] When the Matthean community started proclaiming the gospel, they had serious roadblocks with their fellow Jews, but they had better luck with the Gentiles. So Matthew puts on the lips of the earthly Jesus, "make disciples of all nations," a development that occurred years after Christ's resurrection and ascension. Matthew's gospel is full of references to the Gentiles accepting Christ back to the

---

[27] Donald Senior, *The Gospel of Matthew* (Nashville, TN: Abingdon Press, 1997), 173.

[28] Raymond Brown, *The Church the Apostles Left Behind* (Mahwah, NJ: Paulist Press, 1984), 125.

birth narrative. In the Gospel according to Matthew, Senior notes that "Matthew gives content to the mission of the community. They are to 'make disciples,' that is, form followers of Jesus in the manner affirmed throughout the gospel of Matthew, particularly in the section where Matthew has focused on Jesus' instruction of his disciples."[29] Jesus' instruction to his disciples in the Gospel is still relevant to every disciple today, primarily through the grace each Catholic receives at baptism. Nevertheless, there is an ongoing need to remind the laity of this mandate to engage consciously with inactive Catholics and the unaffiliated.

In elaborating on the mission of making disciples, forming a community is an essential element of the mandate given by Jesus Christ. Senior notes that "Matthew's concern with the formation of the 'Church' was evident in Chapter 16 where Peter was constituted as the 'rock' on which the community was to be built and given authority within the community. The forming of a people who will 'produce the fruits of the kingdom' becomes, therefore, the object of the disciples' mission."[30] Since making disciples is a mandate from Jesus, both clergy and laity need to collaborate to ensure a stable and witnessing community. The instruction to the first disciples to engage and evangelize therefore applies to every Catholic disciple today. The em-

---

[29] Ibid., 176.

[30] Ibid.

phasis on discipleship is also evident when we recognize that the gospel concludes, therefore, with a concentration on Christology, discipleship, and mission—the same concerns that kept pace with the story from the beginning. By ending his narrative with a commissioning text, that is, with the full authority of the risen Christ sending forth the community into its future and with Jesus' promise to remain with the community until the end of time, Matthew ties the narrative world of the gospel within his own community.[31]

The commissioning of the disciples in the Gospel helps us appreciate and embrace our vocation today as an essential aspect of our mission to evangelize within our local parish community.

As one continues to think about the disciple's vocation, as the gospels indicate, it is essential to recognize that it is a vocation that grows and develops in the Christian community. In developing this point, Senior and Stuhlmueller, in their book, *The Biblical Foundation for Mission*, notes that "the mission of the community is rooted in the authority of the risen Jesus who reigns as the Son of God and Son of Man within the community. This foundation is asserted in the opening verses of the pericope.[32] The risen Jesus is ulti-

---

[31] Senior, *The Gospel of Matthew*, 177.

[32] Donald Senior and Caroll Stuhlmueller, *The Biblical Foundation for Mission* (Maryknoll, NY: Orbis Books, 1983), 251.

mately the source of our parish communities. Moreover, every disciple needs to stay connected to the parish community to develop their vocation and prepare for evangelization. As we grow in our vocation as disciples, we begin to appreciate our mission and the need to go out and share our faith. As Senior and Stuhlmueller note, "the first instruction for those sent is that they 'make disciples.' The community is to precipitate the same type of conversation experience that touched the disciples of Jesus: belief in Jesus and transformation of life on that basis."[33] The conversation we have with the inactive or the unaffiliated will involve our experience and encounter with Jesus Christ. Therefore, it is essential to emphasize to members growing as disciples to stay connected to the source of their faith—Jesus Christ and his Church.

A further consideration is that after the ascension of Jesus, the Christian community begins to experience a shift that sends the disciples out of Jerusalem into Gentile territory. The descent of the Holy Spirit and the persecution of the early Christians necessitates a movement away from Jerusalem. In his book, *The Churches the Apostles Left Behind,* Raymond Brown notes that "The first step in making the following of Jesus a missionary movement is attributed by Acts to that Spirit with which the apostles were baptized

---

[33] Ibid., 252.

and empowered to speak."[34] The missionary movement continues to grow as the Acts of Apostles highlight and extends after the conversion of one of the Church's early persecutors—Saul, who the scripture later describes by his Greek name, Paul. The significance of the workings of the Spirit is central in the work of evangelization. The New Testament letters of Paul and the Acts of Apostles highlights this reality.

In thinking of the Holy Spirit's gifts in the life and mission of the Church, the writings from the Pauline epistles give some insight into its distribution and importance in the life of the Church. The apostle Paul in his letter to the Corinthians, says:

> There are different kinds of spiritual gifts but the same Spirit; there are different forms of service but the same Lord; there are different workings but the same God who produces all of them in everyone. To each individual, the manifestation of the Spirit is given for some benefit. To one is given through the Spirit the expression of wisdom; to another the expression of knowledge according to the same Spirit; to another faith by the same Spirit; to another gifts of healing by the one Spirit; to another mighty deeds; to another prophecy; to another dis-

---

[34] Brown, *The Churches the Apostles Left Behind*, 68.

> cernment of spirits; to another varieties of tongues; to another interpretation of tongues. But one and the same Spirit produces all of these, distributing them individually to each person as he wishes (1 Cor 12:4-11).[35]

The gifts, according to Paul, are not for a select group of persons but accessible to all the baptized. Moreover, these individual gifts are for the benefit of the Church and its mission. Since the gifts or charisms are significant factors in evangelization, it will be worth indicating how the New Testament highlights their importance in the Church's mission. Harrington notes:

> According to 1 Corinthians 12:7, every Christian is a gifted or graced person, and has the duty to use those gifts for the good of others:" To each is given the manifestation of the Spirit for the common good." The origin of these gifts is the Holy Spirit. The Christian is the conduit or the instrument of the power of the Risen Christ. The power of Christ made manifest in the present —between Jesus' death and resurrection and the fullness of God's kingdom—is what we call the Holy Spirit. For Paul, each and every Christian is a bearer of the Holy

---

[35] NAB, 1 Cor. 12: 4-11.

> Spirit. The Spirit dwelling within each Christian is the force that expresses itself in the Christian community. And the power of the Holy Spirit is at the root of life in the Christian community.[36]

The gifts are at the root of the Christian life because they animate and inspire the Christian to action. The Christian disciple, therefore, needs to put into action the gifts received. Harrington continues by noting:

> According to Paul, every Christian is a bearer of the Holy Spirit and is led by the Spirit. The actions of individual Christians and of the Christian community are expressions of the Spirit. The various ways in which the activities of the Spirit are expressed are called charisma. The Greek word charisma means "gift." It is related to the term *charis*, which means "grace." To speak of the Church as the "charismatic community" means that it is the assembly of graced persons. "Charism" describes the individuation of Grace, how each Christian responds to grace. The charisms of God's action

---

[36] Harrington, *The Church According to the New Testament*, 58.

> through human persons for the upbuilding of the community of faith.[37]

The apostle Paul, who writes extensively in his letters on the charisms, highlights the gifts' essential nature and importance in the Church's life. The Sacred scripture and the scholarly community both indicate that the gifts each person receives are for a purpose, and that purpose is the building of the community of faith—the Church.

2.3: The Teachings from the Documents and Decrees of Vatican II

In recognizing the divine mandate established in Jesus Christ and the Holy Spirit as witnessed in the New Testament, we turn our attention now to the documents of the Vatican II Council. Though it has been over fifty years since these documents were written and promulgated, the Vatican II documents are the most authoritative Church documents that give us guiding principles for our ministry today. The Council asserts the importance of the laity as collaborators and co-evangelizers in the Church's mission to proclaim the gospel to all nations.

The Second Vatican Council broadened the self-understanding of the Church from a purely hierarchical

---

[37] Ibid., 61.

institution to the people of God, who share a common mission to carry out the work of Christ in the Church and world. Avery Cardinal Dulles, in his book, the *Models of the Church*, notes that "The strong institutionalists development occurred in the late Middle Ages and the Counter-Reformation, when theologians and canonists, responding to attacks on the papacy and hierarchy, accented precisely those features that the adversaries were denying."[38] Therefore, while a response to a crisis at one period in the Church's history, the hierarchy upheld the institutional focus as a standard. Dulles goes on to say:

> In the institutionalist ecclesiology, the powers and functions of the Church are generally divided into three: teaching, sanctifying, and governing. This division of powers leads to further distinctions between the Church teaching and the Church taught, the Church sanctifying and the Church sanctified, the Church governing and the Church governed. In each case, the Church as institution is on the giving end. So these authors say: The Church teaches, sanctifies, and commands, in each case

---

[38]Avery Cardinal Dulles, *Models of the Church* (New York, NY: Image Book, 2014), 28.

> identifying the Church itself with the governing body or hierarchy.[39]

The outlook that Dulles expresses above seems to be prevalent for centuries within the Church. The challenge this presents is that the laity's involvement within the Church gradually diminishes. However, at the turn of the 20th century, the laity's impact becomes more noticeable. At the Vatican II Council, many within the hierarchy acknowledge the growing change and effect. Dulles highlights this point when he states:

> Bishop Emile De Smedt of Bruges spoke of the pyramidal pattern in which all power is conceived as descending from the Pope through the bishops and priests, while at the base, the faithful people play a passive role and seem to have a lower position in the Church. In contrast to this view, the bishop reminded the conciliar Fathers that in the Church all have the same fundamental rights and duties, so that popes and bishops, together with laypersons, are to be reckoned among the faithful people of God.[40]

---

[39] Ibid., 30.

[40] Ibid., 31.

In noting this shift in approach, Dulles indicates how The Council's discussion about the role of the laity evolves. In the ensuing documents of The Council, there was a broadening of an understanding of the laity's role in evangelization and the Church's mission.

Dulles, in his writings, also points out how the institutional ecclesiology had an impact on the discussions of Vatican I. Quoting a debate from the Vatican I schema, he notes that "the Church of Christ is not a community of equals in which all the faithful have the same rights. It is a society of unequals, not only because among the faithful, some are clerics and some are laymen, but particularly because there is in the Church the power from God whereby to some it is given to sanctify, teach, and govern, and to others not."[41] This quote shows that the laity had a lower status within the Church. Such a perception of the laity is bound to impede the Church's mission since it limits their involvement or participation in evangelization.

Dulles and other authors point out that for 500 years prior to Vatican II, the predominant ecclesiology was an institutionalist understanding of the church as a *perfect society*, focused on the powers and role of the hierarchy who exercised the ministry in the Church. At the same time, the laity were passive recipients of their ministry. Vatican II, in

---

[41] Dulles, *Models of the Church*, 31.

many ways, was a response to the institutional focus of the previous 500 years.

The Documents of Vatican II present authoritative teaching that shows the evolution from a hierarchical focus to a focus on the people of God—clergy, religious, and laity. The Dogmatic Constitution on the Church; *Lumen gentium* states:

> The sacred character and organic structure of the priestly community are brought into being through the sacraments and the virtues. Incorporated into the church by Baptism, the faithful are appointed by their baptismal character to Christian religious worship; reborn as sons and daughters of God, they must profess publicly the faith they have received from God through the church.[42]

Through baptism, every member of the Church receives the necessary grace to witness and evangelize others in the faith. Still, it is also the channel whereby both laity and clergy share a joint mission. As one of the sacraments of initia-

---

[42] Vatican Council II, Dogmatic Constitution *Lumen gentium* (21 November 1964) 11, in *Vatican Council II Constitutions and Declarations: A Completely Revised Translation in Inclusive Language,* edited by Austin Flannery (Northport NY: Costello Publishing Co., 1996).

tion, baptism opens the door to all who receive it to be collaborators in the church's mission and prepares them as witnesses of the gospel.

The Council emphasizes the laity's duty to witness to their faith and sees the source of that obligation in the sacrament of baptism. The teaching set out by The Council in this second chapter of *Lumen gentium* shows a broadening in understanding of the Church as the People of God where all have the right and responsibility for the Church's mission to the world. The Council further notes how this prophetic role strengthens the laity in their involvement in evangelization when it notes:

> The holy people of God share also in Christ's prophetic office: it spreads abroad a living witness to him, especially by a life of faith and love and by offering to God a sacrifice of praise, the fruit of lips confessing his name (see Heb 13:15). The whole body of the faithful who have received an anointing which comes from the holy one (see 1 Jn 2:20 and 27) cannot be mistaken in belief.[43]

The prophetic role of the laity, which is my primary focus in the above quote, highlights the living witness of their vocation as a member of the people of God. The exercise of

---

[43] LG 12.

this prophetic role requires participation and proclamation of the good news. The effects of such a proclamation by the laity can open both inactive Catholics and the unaffiliated to a variety of experiences. The document states that "[it] draws its hearers to faith and the profession of faith, it prepares them for Baptism, snatches them from servitude to error, and incorporates them into Christ so that by love they may grow to full maturity in him."[44] The people of God can impact their brothers and sisters through a vigorous response to their baptismal calling. Therefore, "All disciples of Christ are obliged to spread the faith to the best of their ability."[45]

In chapter 2 of *Lumen gentium,* the document provides a broadening concept of the self-understanding of the Church as people of God. Chapter 4 highlights the significance of the laity. The Council describes the laity in chapter 4 when it notes that the term laity is "all the faithful, that is, who by Baptism are incorporated into Christ, are constituted the people of God, who have been made sharers in their own way in the priestly, prophetic and kingly office of Christ and play their part in carrying out the mission of the whole Christian people in the church and in the world."[46] For the first time in official teaching, The Council applies

---

[44] LG 17.

[45] Ibid.

[46] LG 31.

the threefold mission of Christ to the laity. By making this statement, The Council shows support for the laity's vocation and emphasizes their needed participation in the mission of the Church. They clearly assert that the shared mission of Christ was never meant to be the sole mission of the clergy, but rather a collaborative effort of the entire people of God.

> The sacred pastors, indeed, know well how much the laity contributes to the well-being of the whole Church. For they know that they were not established by Christ to undertake by themselves the entire saving mission of the Church to the world. They appreciate, rather, that it is their exalted task to shepherd the faithful and at the same time acknowledge their ministries and charisms so that all in their separate ways, but of one mind, may cooperate in the common task.[47]

As members of the people of God, the fruitfulness of the Church's evangelization mission requires the laity's participation. The Council in *Lumen gentium* further affirms the indispensable role of all the faithful in building up the Church and how their gifts facilitate the fulfillment of the Church's mission in the world. To this point, The Council notes that "It is not only through the sacraments and the ministries of the Church that the Holy Spirit sanctifies and leads the people of God and enriches it with virtues, but,

---

[47] Ibid.

'allotting His gifts according as he wills (cf. 1 Cor. 12:11)', he also distributes special graces among the faithful of every rank."[48] In highlighting the gifts of the Holy Spirit in the Church's mission, The Council provides one more reason to encourage the laity's participation in evangelization.

The Council asserts that the laity is called and gifted to contribute to the Church's mission of evangelization, especially to the local parish community. The document notes that "The laity, however, are given this special vocation: to make the church present and fruitful in those places and circumstances where it is only through them that it can become the salt of the earth. Thus, all lay people, through the gifts which they have received, are at once the witnesses and the living instruments of the mission of the church itself 'according to the measure of Christ's gift' (Eph 4:7)."[49] The Council repeatedly mentions the gifts and the sacraments as essential areas whereby the laity receives strength and power to be channels and instruments of the mediating presence of God to both the inactive Catholic and the unaffiliated in the society. The laity, therefore, has the authority to exercise its prophetic role in the Church and the world with confidence through the grace of Baptism and the other sacraments.

---

[48] LG 12.

[49] LG 33.

The Council recognizes that the collaborative effort of both the clergy and the laity is essential to accomplish the Church's mission of evangelization. In the Decree on the Apostolate of the Laity, *Apostolicam Actuositatem*, the Council emphasizes the indispensable role of the laity in the Church's mission. The Decree notes that "Indeed, the church can never be without the lay apostolate; it is something that derives from the lay person's very vocation as a Christian. Scripture clearly shows how spontaneous and fruitful this activity was in the church's early days (see Acts 11:19-21; 18-26; Rom 16:1-16; Phil 4:3)."[50] The Council traces the laity's involvement back to the early Christian community but indicates how their contribution impacts the entire community today.

If the Church is to achieve its mission in the modern world, it requires an active laity. The demands of the contemporary world differ significantly from times past, in large part due to the proliferation of information technology and the increased engagement arising from the use of social media. To effectively carry out its mission of evangelization, the Church must make use of these technologies.

---

[50] Vatican Council II, Decree on the Apostolate of the Laity *Apostolicam actuositatem* (18 November 1965) 1, in *Vatican Council II Constitutions and Declarations: A Completely Revised Translation in Inclusive Language*, edited by Austin Flannery (Northport NY: Costello Publishing Co., 1996).

The laity also requires proper training and the necessary tools to evangelize our present culture. The need for training and equipping the laity is tremendous, and as the opportunity is present, their formation will open up avenues for engagement.

The ordinary circumstances of the daily lives of the laity provide them with a unique opportunity to engage inactive members or the unaffiliated to a greater extent than the clergy. However, since many members of the laity do not feel confident in engaging inactive Catholics or the unaffiliated, the clergy or religious who have training in evangelization can be resourceful for their formation. As my thesis will note later on the subject of generosity, the laity, like the clergy, are stewards of God's gift. The *Decree on the Apostolate of the Laity* further states that:

> The apostolate is lived in faith, hope, and charity poured out by the Holy Spirit into the hearts of all the members of the church. And the precept of charity, which is the Lord's greatest commandment, urges all Christians to work for the glory of God through the coming of his kingdom and for the communication of eternal life to all, that they may know the only true God and Jesus Christ whom he has sent (see Jn 17:3). On all Christians, accordingly, rests the noble obligation of working to bring all

> people the whole world over to hear and accept the divine message of salvation.[51]

Through the sacraments, each person's gifts prepare and strengthen them to be an instrument of Christ's mission to those within the parish community and those yet to associate with the parish community. The laity also contributes to the Church's mission through the diversity of gifts and charisms they bring to the local parish community and the universal Church. The Decree notes:

> Lay people, too, sharing in the priestly, prophetical, and kingly office of Christ, play their part in the mission of the whole people of God in the church and in the world. In the concrete, their apostolate is exercised when they work to evangelize people and make them holy; it is exercised, too, when they endeavor to have the Gospel spirit permeate and improve the temporal order, going about it in a way that bears clear witness to Christ and helps forward the salvation of humanity.[52]

The expertise of the laity in their professional lives and the gifts they receive in their spiritual lives both contribute

---

[51] AA 3.

[52] AA 2.

to advancing the Church's mission. If the laity participates fully within the parish community and collaborates with their pastors, they can renew the local parish community.

As evangelizers, we need to ensure that the laity receives the necessary instruction on evangelization to promote the faith within or outside our parish community. In keeping with the various challenges and changes within society, the laity as members of the people of God can contribute as long as they remember that the work of conversion and its completion is the Holy Spirit's work. In the Document on the Church in the Modern World, *Gaudium et spes,* The Council notes that "The people of God believe that it is led by the Spirit of the Lord who fills the whole world. Impelled by that faith, they try to discern the true signs of God's presence and purpose in the events, the needs and the desires which it shares with the rest of humanity today."[53] As the Council document highlights, discerning God's will in human activity and society presents some challenges, but with faith, the direction of our work will be made clear through the guidance of the Holy Spirit. The Holy Spirit, who is the source of all gifts, will strengthen our work of

---

[53] Vatican Council II, Pastoral Constitution on the Church in the Modern World *Gaudium et spes* (7 December 1965) 11, in *Vatican Council II Constitutions and Declarations: A Completely Revised Translation in Inclusive Language,* edited by Austin Flannery (Northport NY: Costello Publishing Co., 1996).

evangelization as we strive to accomplish the Church's mission. The Council Fathers, in its *Decree on the Missionary Activity,* adds:

> Therefore, missionaries, the cooperators with God (see 1 Cor 3:9), should raise communities of the faithful, so that walking worthy of the calling to which they have been called (see Eph 4:1) they might carry out the priestly, prophetic, and royal offices entrusted to them by God. In this way, the Christian community will become a sign of God's presence in the world. Through the Eucharistic sacrifice, it goes continually to the Father with Christ, carefully nourished with the word of God it bears witness to Christ, it walks in charity and is enlivened by an apostolic spirit.[54]

Once more, the Holy Spirit's guidance on the affairs of the Church, especially in the activity of the laity, directs them as they witness to Jesus Christ through the grace of the sacraments, especially the Eucharist.

---

[54] Vatican Council II, Decree on the Church's Missionary Activity *Ad gentes* (7 December 1965) 15, in *Vatican Council II Constitutions and Declarations: A Completely Revised Translation in Inclusive Language,* edited by Austin Flannery (Northport NY: Costello Publishing Co., 1996).

To revitalize the Church's mission on evangelization, The Council encourages clergy and laity alike to engage in prayerful reading and diligent study of scripture. In the Dogmatic Constitution on Divine Revelation, *Dei Verbum* notes that "therefore, all clerics, particularly priests of Christ and others who, as deacons or catechists, are officially in the ministry of the word, should immerse themselves in the scriptures by constant spiritual reading and diligent study."[55] The appeal by The Council is about giving scripture its central place in the Church's worship, theology, preaching, and teaching. All of the people of God—clergy, religious, and laity—were to embrace this invitation as a means to nourish and grow in their faith. Accordingly, the Constitution notes that "the holy synod forcefully and specifically exhorts all the Christian faithful, especially those who live the religious life, to learn "the surpassing knowledge of Jesus Christ" (Phil 3:8) by frequent reading of the divine scriptures. 'Ignorance of the scriptures is ignorance of Christ.'"[56] The people of God called to the religious

---

[55] Vatican Council II, Dogmatic Constitution on Divine Revelation *Dei Verbum* (18 November 1965) 25, in *Vatican Council II Constitutions and Declarations: A Completely Revised Translation in Inclusive Language,* edited by Austin Flannery (Northport NY: Costello Publishing Co., 1996).

[56] Ibid.

life were also encouraged to embrace the reading of sacred scripture as a spiritual practice essential for their vocation.

Moreover, The Council further encourages the clergy to provide the proper education and collaborate with the laity in all aspects of the faith to facilitate the Church's mission of evangelization. In its Pastoral Constitution on the Church in the Modern World, *Gaudium et spes*, The Council notes that "those involved in theological studies in seminaries and universities should be eager to cooperate with people versed in other disciplines by pooling their resources and their points of view."[57] While emphasizing the need for this cooperation between the clergy and laity, The Council highlights its importance for the Church's mission. The Decree further notes that "Collaboration of this kind will be beneficial in the formation of sacred ministers; they will be able to present teaching on God, on humanity, and on the world, in a way more suited to our contemporaries, who will then be more ready to accept their word."[58] The Council advocates explicitly for the theological formation of the laity, which they encourage to become professional theologians. The Council's invitation is significant for the laity and women religious who were not permitted to get advanced degrees in theology.

---

[57] GS 62.

[58] Ibid.

In his book on *Rediscovering Vatican II: The Church in the Making*, Richard Gaillardetz sums up The Council's innovative teaching on the laity where he notes:

> The Council taught that the laity have a right and responsibility to be actively involved in the church's apostolate (LG, 30,33). They are equal sharers in the threefold office of Christ who is priest, prophet, and king (LG, 34-36). They are called to full, conscious, and active participation in the liturgy, a participation that is demanded by the very "nature of the liturgy" (SC, 14). Pastors must acknowledge the expertise, competency, and authority of the laity and gratefully accept their counsel (LG, 37). The Council encouraged laypersons to pursue advanced study in theology and scripture (GS, 62). Finally, it is the laity who are to take the initiative in the transformation of the temporal order (LG, 31; GS, 43).[59]

The Council, therefore, promotes active participation of the laity in the Church's mission to evangelize the Church and the world about the love of God for us.

---

[59]Gaillardetz, *Rediscovering Vatican II: The Church in the Making* (Mahwah, NJ: Paulist Press, 2006), 53.

Hence, the laity has a significant role in the Church, especially today, as the Church finds ways to engage inactive and non-Catholic persons. I will add that in a rural community where the presence of the Catholic Church is not strong, there is a need to collaborate with the laity to develop ways that will engage inactive Catholics and the unaffiliated.

2.4: Papal Encyclicals and Exhortations

As the preceding section notes, many Vatican II documents highlight the laity's crucial contributions to the Church's mission in the world. Likewise, the United States Conference of Catholic Bishops emphasizes that the laity's involvement in the mission of the Church is essential. In its document, *Co-Workers in the Vineyard of the Lord,* the United States Bishops indicate that "the Church is the communion of those called by Christ to be his disciples. Discipleship is the fundamental vocation in which the Church's mission and ministry find full meaning. The call to discipleship is, first of all, a gift. God offers to us a share in the Trinitarian communion, the love of Father, Son, and Holy Spirit."[60] Since we are all disciples and members of the

---

[60] United States Conference of Catholic Bishops, *Co-Workers in the Vineyard of the Lord: A Resource for Guiding the Develop-*

body of Christ, working together to accomplish our mission needs to be a priority for both clergy and laity. The statement by the bishops goes on to say:

> Ministry is diverse and, at the same time, profoundly relational. This is so because ministry has its source in the triune God and because it takes shape within the Church understood as a communion. Ministerial relationships are grounded first in what all members of Christ's Body have in common. Through their sacramental initiation all are established in a personal relationship with Christ and in a network of relationships within the communion of the People of God. The personal discipleship of each individual makes possible a community of disciples formed by and for the mission of Christ.[61]

Since all God's people share a common baptism, it is necessary to engage and evangelize from that common ground. Notwithstanding, personal discipleship, as the bishops note, is vital in fostering communal worship. If the laity recognizes their duty and receives formation to

---

ment of Lay Ecclesial Ministry (Washington D.C.: United States Conference of Catholic Bishops, 2008), 19.

[61] Ibid., 21.

strengthen their state, the Church will benefit from the fruits of such endeavors.

Pope Paul VI, in his writing at the end of Vatican II Council, affirms the necessity of evangelizing and its importance. In his apostolic exhortation on Evangelization in the Modern World, *Evangelii nuntiandi,* Pope Paul VI states:

> The Church is an evangelizer, but she begins by being evangelized herself. She is the community of believers, the community of hope lived and communicated, the community of brotherly love, and she needs to listen unceasingly to what she must believe, to her reasons for hoping, to the new commandment of love. She is the People of God immersed in the world, and often tempted by idols, and she always needs to hear the proclamation of the "mighty works of God" which converted her to the Lord; she always needs to be called together afresh by Him and reunited.[62]

Thus, the task of evangelizing begins with attentive listening and encounters with the word of God, Jesus

[62] Paul VI, Apostolic Exhortation *Evangelii nuntiandi* (8 December 1975), (Frederick, MD: The Word Among Us Press, 1975) 15.

Christ. Just as food is necessary for growth and development in physical life, the Word of God is essential for growth and development in individuals' spiritual and moral lives. Our relationship with God and others can quickly grow stale if not sustained adequately through prayer and an ongoing encounter with the Word of God, among many other ways provided to us by the Church.

Reaching out to inactive members and, when possible, the unaffiliated are equally essential in evangelization. Building trust and finding a connection with this group—the inactive and the unaffiliated—may be challenging but not impossible. As Pope John Paul II in *Redemptoris missio* notes, "Today the Church must face other challenges and push forward to new frontiers, both in the initial mission *ad gentes* and in the new evangelization of those peoples who have already heard Christ proclaimed."[63] The mission to all people, however, begins with our primary mission to our local parish community. To this point, Pope John Paul II notes that "people today put more trust in witnesses than in teachers, in experience than in teaching, and in life and action than in theories. The witness of a Christian life is the first and irreplaceable form of mission: Christ, whose mission we continue, is the 'witness' *par ex-*

---

[63]John Paul II, Encyclical Letter *Redemptoris missio* (7 December 1990), (Washington D.C.: United States Conference of Catholic Bishops, 1990) 30.

cellence (Rv 1:5; 3:14) and the model of all Christian witness."[64] In expressing the importance of our call to evangelize, the Pope stresses that as witnesses, we are also missionaries whose proclamation strengthens the parish community's life.

Therefore, there is a continuous invitation to the laity to participate in evangelization in our time. The work of evangelization carried out by all the members of the Church can facilitate a renewal within every parish and diocese. In line with this desire for renewal, Pope Francis in *Evangelii gaudium* states:

> In this preaching, which is always respectful and gentle, the first step is personal dialogue, when the other person speaks and shares his or her joys, hopes, and concerns for loved ones or so many other heartfelt needs. Only afterward is it possible to bring up God's word, perhaps by reading a Bible verse or relating a story, but always keeping in mind the fundamental message: the personal love of God who became man, who gave himself up for us, who

---

[64]RM 42.

is living and who offers us his salvation and his friendship.[65]

Through personal dialogue, relationships and a sense of trust that allows evangelization can develop within our communities. While it may not seem like a new way to evangelize, this style of evangelization enables us to connect with our listeners. The personal dialogue is what creates an atmosphere that leads to faith sharing.

Our missionary call is part of our vocation as Christians. In accord with our vocation as Christians, Pope Francis in *Evangelii gaudium* writes that "Every Christian is a missionary to the extent that he or she has encountered the love of God in Christ Jesus: we no longer say that we are 'disciples' and 'missionaries,' but rather that we are always 'missionary disciples.'"[66] Discipleship challenges us to evangelize outside the boundaries of the Church and reminds us to seek out all who have stepped away from active participation in the Church. It is also noteworthy that discipleship is a daily and ongoing process for all Christians.

---

[65] Francis, Apostolic exhortation *Evangelii gaudium* (24 November 2013), (Washington D.C.: United States Conference of Catholic Bishops, 2013) 128.

[66] EG 120.

2. 5: Implications for Evangelization

In looking at the implications of the biblical and theological foundations for my work, I see how broadening the laity's involvement contributes to evangelization in various ways. These insights and instructions can guide pastors and their pastoral teams in developing formation programs that provide a collaborative framework for evangelization. These programs can engage and prepare the laity to evangelize inactive Catholics and the unaffiliated. In many statements and documents, the Catholic Bishops of the United States continue to sensitize the laity to participate in ways to ensure that everyone contributes to evangelization. The Bishops' Committee on Evangelization and Catechesis reiterate the mandate by Jesus when it notes in its document, *Go and Make Disciples,* that "We must evangelize because the Lord Jesus commanded us to do so. He gave the Church the unending task of evangelizing as a restless power, to stir and to stimulate all its actions until all nations have heard His Good News and until every person has become a disciple."[67] As with the mandate to make disciples, the bishops stress the need for community. Therefore, the presence of a

---

[67] Committee on Evangelization and Catechesis, United States Conference of Catholic Bishops, *Go and Make Disciples: A National Plan and Strategy for Catholic Evangelization in the United States* (Washington D.C: USCCB Inc. 2002), 17.

community facilitates the growth of each person's faith and their encounter with God.

Moreover, when speaking of the formation of a community, the training of the individual disciple is equally essential. Solid communities are often a result of reliable and dedicated disciples. Hence, targeting the disciples' faith formation for evangelization and mission is necessary. The disciple's faith formation in our rural parish community is an essential part of our process to engage inactive Catholics and the unaffiliated. If making disciples is a primary consideration for community formation, it is necessary to ensure that all the baptized recognize and appreciate their vocation as disciples. The mandate by Jesus to his disciples reminds us that the disciples must engage in evangelization for the Church's mission to be fulfilled. The commissioning, which aligns with the departure of Jesus, places the Church's mission in the hands of the present disciples but hopes that future disciples will embrace this mission as essential to the living out of their faith.

In renewing our parish community, our learning process will involve a training program instructing and preparing members for evangelization. Both The Council Documents and the Papal teachings and exhortations stress the significance of evangelization and the need for formation. The daily encounter with the Word of God through scripture reading and meditation is equally crucial in forming the laity and the development of their faith. These elements highlight the process that serves as reminders to my local parish community. As the parish anticipates engaging inactive Catholics and the unaffiliated within the Southeast region of the State of Missouri, these considerations stand out as crucial for its formation.

Over time, due to a lack of motivation or confidence in themselves, the desire to evangelize or share their faith fades away or becomes dormant. Through a process of formation and training, the laity within my parish community will have the opportunity to develop skills that will facilitate their engagement of inactive Catholics and the unaffiliated.

Evangelization in every context may require a different approach; however, the expectation that the evangelizer be grounded in their faith is essential. Within our rural parish community, one way we prepare ourselves is through catechesis and faith formation programs. Formation and catechesis throughout evangelization equip members with the knowledge necessary to carry out their mission. The United States Conference of Catholic Bishops highlights this when it notes in its document, *Co-Workers in the Vineyard of the Lord,* that "Some mid-life adults, with significant volunteer ministerial experiences in their background, may develop a desire to serve the Church in a fuller way. Retired adults, recognizing the spiritual potential of their accumulated experience and knowledge, personal financial resources, and time, find themselves free to follow desires they had put aside to fulfill family or professional obligations."[68] Therefore, the strength of our formation will be foundational for members who desire to work as witnesses in evangelizing and inviting inactive members to active participation in the Church's life. In other words, this formative step will provide all parish community members a twofold goal of renewal and preparation. In engaging in such a formative process that renews and prepares, members become more established as disciples.

The Bishops of the United States, in their document on a strategy for evangelization, *Go and Make Disciples* note the following possible approaches in the Sunday Eucharist, namely:

- Greeting and welcoming of people;
- Creation of a greater sense of prayer during Mass;
- More clear calling to worshipers to conversion and renewal;
- Fostering a sense of community among parish members;
- Outreach to visitors and newcomers at Sunday Mass;
- Accessibility to worship for everyone.[69]

---

[68]USCCB, *Co-Workers in the Vineyard of the Lord*, 27.

[69] United States Conference of Catholic Bishops, *Go and Make Disciples: National Plan and Strategy for Catholic Evangeli-*

These strategies can engage all parish community members and provide an opportunity to welcome and include either returning inactive members or the unaffiliated within the parish community. As our parish community looks ahead to evangelizing, some of the suggestions from the United States Conference of Catholic Bishops will be essential for our process and program. The bishops further note that "the goals and strategies of our plan are not meant to be an added burden on already overworked pastoral staffs, as if evangelization were merely another program to be done. Rather, they should help parishes see the evangelizing potential of their current activities, even as they stretch parishes to develop new activities from a renewed spiritual energy."[70] These appeals and directives will ground my intervention and thesis as I develop a training program for evangelization. The focus on the laity creates an awareness that needs some emphasis today as the Church in Southern Missouri, like in other parts of the country, sees a decrease in active participation. As Gaillardetz notes, "the life witness of the layperson is but a more intensive realization of the evangelical imperative incumbent upon all Christians to give testimony of God's saving love in the

---

zation in the United States (Washington D.C.: United States Conference of Catholic Bishops, Inc., 2002), 51.

[70] Ibid., 47.

world."[71] As a pastor, in collaborating with the members of my parish community, I see the need to invite and challenge them to engage inactive members and, when necessary, reach out to the unaffiliated.

However, the gifts every baptized person receives through the sacraments for the benefit of the Church play a significant role when we think of evangelization. In his book, *The Church and Ecumenism,* Hans Küng writes that "This sharing of all the People of God in Christ's prophetic office means simply that the whole people, communally and individually, is "anointed," filled and moved by the Holy Spirit."[72] Being filled with the Holy Spirit, therefore, strengthens every baptized Christian with gifts for the mission. Küng goes on to say

> The quoted texts about the whole Church being filled with the Spirit clearly affirm that every Christian is directly enlightened by the Spirit, and this plainly has priority over any human enlightenment: "But you have the anointing that comes from the holy one, and you know all things . . . And As for

---

[71]Gaillardetz, *Rediscovering Vatican II*, 55.

[72] Hans Küng, ed., "The Charismatic structure of the Church," in *Ecumenical Theology: The Church and Ecumenism*, vol. 4, *Concilium: Theology in the Age of Renewal*(New York, NY: Paulist Press, 1965), 42.

> you, let the anointing which you received from him dwell in you, and you have no need that anyone teach you. But as his anointing teaches you concerning all things, and is true and is no lie, even as it has taught you, abide in him." (1 John 2, 20. 27).[73]

The gifts of the Holy Spirit through Baptism and Confirmation are not for a select few within the Church but all Catholics. To highlight this fact, Küng notes:

> The New Testament shows that it is wholly impossible to limit the charismata to officeholders. To start with, we should realize that Paul answers the question about the division and ordering of the "hierarchy" very differently than is customary today. He emphatically ranges them as follows: "And God indeed placed some in the Church, first as apostles, secondly (other ones) prophets, thirdly (still others) teachers; after that miracles, then gifts of healing, services of help, power of administration, and the speaking of various tongues . . . Yet strive after the greater gifts." (1 Cor. 12, 28-31).[74]

---

[73] Ibid., 43.

[74] Ibid., 55.

The gifts that every baptized person receives highlight why the laity needs to participate fully in the mission of evangelization in the Church. Brown elaborates on the significance of the Holy Spirit in the life of the Church when he states that

> The Spirit that brought faith to the Gentiles and brought Paul to Rome continues and will help the church in moments of need. For Christian self-understanding, how important through the centuries has been the idea that the Holy Spirit will not let the Church down! When Christians encountered error and stupidity that seemed to threaten survival, how often have they exclaimed, "Thank God there is the Holy Spirit to pull us through, despite Church leadership." Again and again in church history, when something marvelously unexpected happened, Christian Faith has discovered there the Spirit guiding the Church.[75]

Küng highlights that within the Church today, every Christian disciple has the ability through the gifts and workings of the Holy Spirit to be a part of the mission of the Church. No one has any excuse not to participate in the building up of the body of Christ, which is the Church. The

---

[75] Brown, *The Churches the Apostles Left Behind*, 70.

diversity of charisms within the parish community facilitates Küteamwork, and this can enhance evangelization.

2.6: Conclusion

In looking ahead, the biblical foundation, the Church's teaching, the writings of the Popes, and the statements from the United States Conference of Catholic Bishops all have one thing in common: evangelization is part of our mission as Catholics. Michael Monshau, in his edited work on, *The Responsibility of Every Christian: To Preach Jesus Christ*, Monshau writes that "the contemporary Church, however, has gone well beyond restoring preaching to its rightful role at worship; the Church also insists today that all members of the Church must actually participate in that all-important preaching mission!"[76] In essence, the Church affirms all its members to participate in her mission of spreading the Word of God, and evangelization is part of that preaching process. The next chapter of this thesis will highlight the leading lay evangelists whose resources and writings contribute to my ministerial intervention. I will

[76] Michael Monshau, "The Responsibility of Every Christian: To Preach Jesus Christ" in *Listening: Journal of Religion and Culture,* Mark McVann, F.S.C. ed., vol. 38, No. 1, Winter 2003,The Catholic Recovery of Preaching(Ann Arbor, MI: Lewis University, 2002), 23.

describe the essential aspects of their writings that contribute to my interventions for evangelization.

Chapter 3
Preaching Beyond the Pulpit

3.1: Introduction

In my last chapter, I discussed the biblical and theological foundation of my thesis project and the indispensable role of the laity in the Church's mission of evangelization. In this chapter, I turn my attention to the lay evangelists whose resources on evangelization will guide the ministerial intervention of my thesis project. The leading lay evangelists in this research are Matthew Kelly and Sherry Weddell. Both Kelly and Weddell have written and worked extensively on lay evangelization, albeit from different perspectives.

As I reviewed the field of Catholic evangelization in search of the evangelical methods best suited for my ministerial context in rural southeast Missouri, I read the writings and engaged the resources of some of the leading United States Catholic evangelists, all of whom happened to be laypeople. I will begin by sharing the fruits of that research with a discussion of the following lay evangelists: Curtis Martin, Scott Hahn, Ralph Martin, Matthew Kelly, and Sherry Weddell. They each have a unique approach to evangelization partly because of their different target audiences and methodologies. A brief review of these lay evangelists will help the reader in understanding my choice of Kelly and Weddell. Next, I will focus on the two lay evangelists who have most influenced the ministerial intervention of my thesis project: Matthew Kelly and Sherry Weddell. I will, first, discuss Matthew Kelly's contributions to lay evangelization and his four signs of a dynamic Catholic. I will then highlight Sherry Weddell's contribution to lay evangelization and her *Called and Gifted* discernment process. Throughout my discussion, I will draw out the contribution each lay evangelist makes to my thesis project on parish evangelization in rural Missouri. At the end of this chapter, I will give a brief conclusion and preview for the next chapter.

3.2: Lay Evangelists in the Field of Evangelization

I start with Curtis Martin, the president and founder of the Fellowship of Catholic University Students (FOCUS).[77] He founded FOCUS in 1998 to engage and evangelize college students in their faith. Besides his work with FOCUS, Martin is a Consultor to the Pontifical Council for Promot-

---

[77] Curtis Martin, *The Real Story: Understanding the Big Picture of the Bible* (North Palm Beach, FL: Wellspring), 153.

ing New Evangelization.[78] His work on evangelization is quite popular on college campuses; his approach aligns more to a younger audience and focuses on college-aged men and women away from home and in need of a renewal of their faith as they navigate the college atmosphere.

In his work, Martin challenges the FOCUS missionaries to engage the faith of other students in an intentional manner within the context of the university. He is edified to see young men and women living within the university's culture have the courage to share their faith. If young adults can share their faith with other young adults, I believe that we can strive to imitate their boldness and intentional approach in sharing faith within our context in my rural parish community. In his book on *Making Missionary Disciples*, Martin notes:

> Working in an environment that could be called one of the most hostile to religion and culture—the university campus—FOCUS has seen the Church flourish and grow. While programs are vitally important, our work has confirmed, over and over, that people are everything. Everything begins to change when a culture of missionary disciples is es-

---

[78] Curtis Martin, *Making Missionary Disciples: How to Live the Method Modeled by the Master* (Genesee, CO: FOCUS, 2018), 64.

> tablished (cf. Mt 5:13-16). As the saying goes, "Culture eats strategy for breakfast." When a plan to foster missionary disciples is implemented in your parish, apostolate, or family, lives are changed, churches come alive, and poverty in all its forms begins to be addressed.[79]

From Martin's perspective, having a plan in place is always a good place to start. As a pastor, I think his recommendation to have a plan to evangelize is a first step in the entire process.

Another aspect of the FOCUS missionaries' strategy that may be helpful for parochial work is the idea of inviting outside experts to help evaluate our parish strategy. Martin continues by noting that "After consulting with our missionaries, evaluating our inputs and outcomes, and even hiring an outside researcher to evaluate the long-term impact of our efforts, we have grown in our understanding of the key habits in our missionaries that helped them be more effective in evangelization and the approaches they took that had the biggest impact for raising up missionary disciples."[80] Martin describes the key habits for effective mis-

---

[79] Ibid., 4.

[80] Ibid., 6.

sionary work as divine intimacy, authentic friendship, and clarity and conviction about spiritual multiplication.[81]

The key habits are worth highlighting since they are effective practices that FOCUS missionaries use to engage other students within the university. Elaborating on these key habits, Martin notes that, "Evangelization is first and foremost the work of God, and we will be fruitful in the mission of sharing the gospel only to the extent that we ourselves are abiding in deep union with him"[82] Therefore, intimacy with God through prayer is essential for evangelization. Martin further notes that "we must genuinely love the people we are serving, accompanying them in life and personally investing ourselves in them with authentic friendship"[83] In highlighting the need for authentic friendship, Martin invites every FOCUS missionary to know the students they engage as a way of building trust. In describing the third habit of Clarity and Conviction, Martin notes that "[it] is that laser focus which keeps us on task: *clarity* that is the most effective way to fulfill the Great Commission to make disciples of all nations (Mt 28:19), and *conviction* that we must prioritize our actions to make time for these relationships"[84] The goal of this third habit is to form

---

[81] Ibid.

[82] Ibid., 10.

[83] Ibid., 17.

[84] Ibid., 22.

a small group of missionaries who will, in turn, go out and engage other students as missionaries to invite them to be missionaries. Martin also notes that "spiritual multiplication is exemplified by Jesus, who invested deeply in a remarkably few disciples and trained them to do the same so that the whole world could be reached with the Gospel."[85] These habits, according to Martin, have been tested and are effective for the FOCUS missionaries.

From Martin's methodology, I will engage my parishioners in spiritual multiplication by initiating a day of recollection at the start of each liturgical season as an opportunity for members to learn about specific aspects of the Catholic faith. When parishioners feel comfortable about their faith, they will be more willing to share their faith with other parishioners. Such formation opportunities are what this project will facilitate.

Scott Hahn, a former Presbyterian, is a convert to Catholicism and the founder and president of St. Paul Center for Biblical Theology.[86] Besides being an author of numerous books, he is a professor at the Franciscan University of Steubenville. He holds the Father Michael Scanlan Chair of

---

[85] Ibid.

[86] Scott Hahn, *Evangelizing Catholics: A Mission Manual for the New Evangelization* (Huntington, IN: Our Sunday Visitor Publishing, 2014), 177.

Biblical Theology and the New Evangelization.[87] His academic background makes his work deeply theological, and this may be in some ways challenging for a rural community who desires something simple for their faith formation and preparation for evangelization. In his evangelical strategy, Dr. Hahn, whose protestant background influences his approach to sharing the faith, challenges Catholics to be confident in sharing their faith. He also encourages ongoing studies about the faith to prepare Catholics to respond to questions about the faith. In his book, *Evangelizing Catholics: A Mission Manual for the New Evangelization*, Dr. Hahn notes:

> Not long ago, I put a question to the students in my graduate theology course on the New Evangelization. "How many of you have brought a person to faith in Christ, from unbelief to baptism?" I asked. Two hands went up. Out of forty students. And one of those hands belonged to a former missionary. Admittedly, it was a question I shouldn't have asked. It had a distinctly Protestant flavor to it and, in many ways, missed the point of the New Evangelization, which is about so much more than racking up numbers of souls saved. Nevertheless, the results of that quick survey pointed to one of the New

---

[87] Ibid.

> Evangelization's primary obstacles – namely, that Catholics don't have much experience when it comes to sharing their faith with others. Many are reluctant to do it. So, many don't do it.[88]

As a convert to Catholicism, Dr. Hahn shows that sharing the faith does not seem to be a priority for many Catholics, unlike many mainline Protestant denominations. However, he challenges his students and every Catholic to see this form of evangelization as sometimes necessary through his approach. Dr. Hahn further notes:

> When you know Jesus – when you know who he is, what he's done for you, what he's sacrificed for you, what he's given you – it's hard not to be overjoyed. It's also hard not to share all you know with others. You want to tell people about Jesus. You want to tell people about his Church. You want to tell them about the teachings, and the graces and the saints and the sacraments and everything you've encountered that has changed who you are. You can't help it. It just flows out of you.[89]

---

[88] Ibid., 27.

[89] Ibid., 29.

For Dr. Hahn, knowing one's faith and sharing one's faith needs to go together as one grows. In the *Bible belt*, where Catholics are a small percentage, it may be a tall order, but one may challenge Catholics to greater involvement in learning and knowing about their faith.

From Dr. Hahn's methodology, developing a regular bible study class for the members of my parish would be significant. The classes will introduce members to the scriptures and help those proficient in the scriptures support other members who may need guidance on the biblical story. In forming the laity, an understanding of scripture is essential for evangelization and vital for our faith.

Ralph Martin is another lay religious teacher whose work provides some resources for evangelization. He dedicates his time to fostering renewal and evangelization in the Catholic Church and is president of Renewal Ministries.[90] His work on evangelization focuses on helping Catholics, both clergy and laity, deepen their spiritual lives. His background on the spiritual life is a helpful resource for Catholics who desire to deepen their relationship with God. While his work will be beneficial at an advanced level when members develop a program, his style does not provide a

---

[90] Ralph Martin, *The Urgency of the New Evangelization: Answering the Call* (Huntington, IN: Our Sunday Visitor Publishing, 2013), 144.

rural context with resources to facilitate an initial program for evangelization.

In his writings, Martin advocates engaging the secular culture and forming Catholics to have the tools to engage the culture. In his book *The Urgency of the New Evangelization*, Martin notes the following:

> Many reasons are given why it is important that there be a new evangelization. Chief among them seems to be the sober realization that traditionally Christian and Catholic countries are now under powerful assault by secularist and even anti-Christian forces. As a result, many millions of our fellow believers may still have the name of Christian or Catholic but are no longer living as disciples of Christ or even understanding any more what it would mean to do so. This shows up not only in declining Church attendance but in the erosion of belief and morality even among those who still do attend. Other reasons that are given to demonstrate the need for a new evangelization include theological reasons that show that evangelization is the primary mission of the Church and that we have been explicitly commanded by Christ to evangelize. Mention is also frequently made about how becoming a Christian fulfills human nature, gives us eternal life, and brings us to salvation, rescuing us from sin,

> with an occasional mention of Satan as well. But even though these words are spoken and written, they often fail to "connect" with any sense of urgency because of a prevailing mentality that dulls our perception of the urgency for evangelization.[91]

Martin's invitation reminds us that fundamental aspects of the Catholic faith, foundational for our development as Catholics, are experiencing a significant crisis. Therefore, he challenges Catholics to look within these essential aspects of our faith as we engage the culture.

Furthermore, Martin reminds lay evangelists of the importance of the scriptural and theological background for evangelization. He writes that "According to Matthew, disciples are those whose identity is founded in their relationship with Jesus and who make him the absolute priority of their lives (Mt 8:19-23; 10:37-39; 16:24-26). They share in Jesus' mission his authority, his self-denial, his persecution, and his reward (Mt 10; 19:29-30)."[92] For Martin, our relationship with Jesus Christ strengthens our identity as disciples and enables us to evangelize the inactive or the unaffiliated.

From Martin's evangelization work, helping parishioners develop a love for the sacraments, especially the Eucha-

---

[91] Ibid., 16.

[92] Ibid., 112.

rist and reconciliation, will be essential. In a climate where participation in weekly Eucharistic celebration and reconciliation in certain dioceses is declining, a class on the importance of these sacraments and all the sacraments will be necessary.

As I look at each of the above lay evangelists, it is evident that they all emphasize the importance of a personal relationship with Jesus Christ as a requirement for evangelization. Kelly and Weddell's approach to evangelization has this in common and, in their resources, offer some practical ways to develop a habit of prayer in the lives of the laity. Nevertheless, each person's approach to outlining a program or process that prepares the laity to engage either the inactive Catholic or the unaffiliated is different. Moreover, Kelly and Weddell, as we will see, highlight simple programs that can provide a starting point for the development and training of the laity within a parish community. In choosing resources from the work of Kelly and Weddell, I will engage my parishioners by enabling them to embrace a process that will help them grow in their appreciation of evangelization.

3.3: Matthew Kelly's Contributions to Lay Evangelization

Matthew Kelly is the founder of the Dynamic Catholic Institute with the mission of "re-energizing the Catholic Church in America by developing world-class resources to inspire people to rediscover the beauty of the Church."[93] His evangelization approach involves engaging members within the parish community to see their faith's beauty and depth. He does this by exploring an encounter with the four signs of a dynamic Catholic: prayer, study, generosity, and evangelization.[94]

Kelly shares how this journey started for him. He writes how an experience at a dinner table triggered his journey. He overheard an older priest at the table responding to a young priest's question by saying, "I have been the pastor of seven parishes over the past forty years, and I can tell you that it doesn't matter where you go, you will discover the same fifty people do everything in a parish."[95] This response was the beginning of Kelly's involvement in the work of Parish evangelization. He continues by noting that "in the following weeks, I started making informal phone calls to some pastors I knew. I asked them questions like:

- Who are your most engaged parishioners?
- Why are they so engaged?
- What percentage of registered parishioners are actively involved in the parish?

---

[93] Matthew Kelly, *The Four Signs of a Dynamic Catholic* (Hebron, KY: Beacon Publishing, 2012), 223.

[94] Ibid., 18.

[95] Ibid., 11.

- What percentage of the parishioners regularly give to the parish."[96]

He organized these questions and their responses into data for his work on evangelization, which he continues to use as part of his study. These questions facilitated his initiative to begin the Dynamic Catholic Institute. However, he points to the Vatican II document—*Lumen gentium*, as foundational for his work. He highlights the universal call to holiness as essential to the life of all the baptized.[97]

In his work on evangelization, Kelly highlights how dynamic Catholics have a basic routine of learning. He describes dynamic Catholics as "highly engaged parishioners."[98] The daily or weekly routine will determine their commitment to the faith or a need to grow as a disciple. In his approach, he challenges every Catholic to desire to grow in their faith, indicating that "the things we do repeatedly determine our character and destiny. This is equally true for an athlete, a business leader, a parent, or a Catholic."[99] In other words, who we are is connected to what we do. For

---

[96] Ibid.

[97] Matthew Kelly, *Rediscover Catholicism: A Spiritual Guide to Living with Passion and Purpose* (Hebron, KY: Beacon Publishing, 2010), 70.

[98] Kelly, *The Four Signs of a Dynamic Catholic,* 13.

[99] Ibid., 16.

his research, he uses four direct components to help struggling parishes, namely:

- Catholic Resources: books, CDs, events, retreats, and pilgrimages
- Consulting: a dedicated consultant provides and helps in the implementation of parish initiatives
- Executive Coaching twice a month for the pastor and leadership team
- Research through "The Bigger Future Survey" to check engagement levels and progress throughout five years.[100]

In 2018 the Dynamic Catholic Institute began partnering with 21 parishes across four dioceses. The Dynamic Catholic program incorporates activities that engage in catechesis for adults and children. These programs include "*Starting Point*" for Baptism, "*Blessed*"—for first communion and reconciliation, "*Decision Point*" for confirmation, "*Better Together*" for marriage preparation and enrichment. Other programs include "*Welcome CRHP 2.0,*" a parish retreat program, the "*Best Lent Ever*" for the season of Lent, the "*Best Advent Ever*" for the advent season, and a "*Parish*

---

[100] Matthew Kelly, *Dynamic Catholic Programs,* Web post accessed 10 September 2020 at https://d2y1pz2y630308. cloudfront.net/2415/documents/2019/12/DynamicParish_Invite.pdf

Book Program" that encourages ongoing study to nourish and nurture one's faith.[101]

Moreover, since his program incorporates a revised version of Christ Renews his Parish, Kelly notes that "in 2017, we rebranded CRHP as '*Welcome*' and launched a pilot [program] in 175 parishes throughout the country. We worked closely with 21 parishes who had never used CRHP before and over 150 parishes who had already been using CRHP to introduce the updated program and materials. The pilot will wrap up in 2018, setting the stage for a nationwide launch."[102] Since the conclusion of these programs, it has been two years now, and the Dynamic Catholic Institute continues to expand its outreach to other parish communities. Nevertheless, in an earlier study in preparation for this pilot program, Kelly notes that "over many months I studied a series of parishes from coast to coast, examining two areas in particular: volunteerism and financial contribution. Both are significant signs of engagement."[103] And his discovery amazed him because it showed the following:

---

[101]Ibid.

[102] Matthew Kelly, *Dynamic Catholic Programs* Web post accessed September 10, 2020 [1:50 p.m.], accessed online September 10, 2020, https://greatnonprofits.org/org/dynamic-catholic-institute.

[103]Kelly, *The Four Signs of a Dynamic Catholic*, 12.

- 6.4 percent of registered parishioners contribute 80 percent of the volunteer hours in a parish.
- 6.8 percent of registered parishioners donate 80 percent of financial contributions
- There is an 84 percent overlap between the two groups.[104]

In his research with the Dynamic Catholic Institute, Kelly notes that "roughly 7 percent of Catholic parishioners are doing almost everything in their faith community and paying almost entirely for the maintenance and mission of the parish."[105] The four signs challenge the 93 percent to work towards greater involvement in the Church's evangelization mission. As a strategy from his work on *Dynamic Catholic*, I will incorporate in my intervention the four signs as a means to train and engage the participants in this research as we work to evangelize the inactive Catholics within our parish community. The four signs of prayer, study, generosity, and evangelization complement each other, especially concerning evangelization. As part of my strategy, all participants will journal where they are at the start of the training and continue to journal their progress. The journaling practice will allow each person to determine

---

[104] Ibid.

[105] Ibid.

their progress throughout the interventions and testing phases. Kelly's approach is interpersonal, and he does not mention any journaling. However, since this research will engage Catholics and the unaffiliated through social media, keeping track of the various changes throughout the process will enhance proper data collection. The use of personal journals will provide participants with a tool for gauging their progress and an opportunity to share their journey of growth with other participants.

3.4: Kelly's Four Signs of a Dynamic Disciple

I have chosen to use the four signs recommended by Kelly as part of my intervention because it will challenge all participants in this research to realize that the work of evangelization requires ongoing personal growth in prayer and study. However, through our engagement and development in prayer and study, the desire to be generous as we work to evangelize ourselves and other active and inactive Catholics are likely to increase.

Regarding prayer, Kelly notes that it determines if one is spiritually healthy.[106] The spiritual health of every evangelizer is essential to the work of evangelization. Also, determining each participant's spiritual temperature, I will develop a ten-minute rule which will require making a ten-

---

[106] Ibid., 41.

minute sacrifice each day of personal prayer. I will use this sign as part of my strategy because it will contribute to growth in each person's relationship with God.

When we think of study, Kelly notes that "study fuels prayer, generosity, and evangelization. The more we learn about our faith, the more likely we are to take the spiritual life seriously."[107] The study strategy will focus on weekly bible study and catechesis on topics that deal with nurturing the faith like the sacraments, the saints, and areas within the Catechism of the Catholic Church that will prepare active Catholics to engage inactive members and non-Catholics. These sessions will involve listening to scholars like Dr. Scott Hahn and Bishop Robert Barron. They have excellent recommendations that can help the laity in building their interest and knowledge of Church teaching. I have chosen this approach because it will give the participants materials for discussion and a new opportunity to engage in questions that will challenge them to grow and work towards evangelization.

In thinking of generosity, Kelly notes that "generosity is the trademark of Dynamic Catholics. They are generous with their time and talent and with their money and possessions, but their generosity goes way beyond these commonly defined areas."[108] Kelly's premise on generosity highlights

---

[107] Ibid., 94.

[108] Ibid., 139.

that our generosity as Catholics is not limited to money but extends to other areas of our faith, including but not limited to service within the Church and volunteering for various ministries. As regards generosity, this will impact my strategy because it will help highlight some of the gifts each participant brings to the work of evangelization. Kelly's approach to generosity as a sign allows me to incorporate the recommendations from Sherry Weddell on the use of charisms. I will draw on Weddell's spiritual gifts inventory here to help participants realize that generosity extends to other areas of our life as Catholics, like hospitality, teaching, writing, service, and so forth.[109]This will be an opportunity for members to discern some of their gifts as they prepare for the last phase of evangelization. More will be said in the area of generosity when we look at the life and writings of Sherry Weddell and the Catherine of Siena Institute, of which she is a co-founder.

The last sign of a dynamic Catholic proposed by Kelly is that of evangelization. In this phase of discipleship, the dynamic Catholic learns to approach evangelization from what Kelly calls a "win, build and send" model.[110] Kelly explains that "God wants to *win* you with his love and wisdom; God wants to *build* you up spiritually so that you have

---

[109] Sherry Weddell, *The Catholic Spiritual Gifts Inventory* (Colorado Springs, CO: The Siena Institute Press, 1998), 60.

[110] Kelly, *The Four Signs of a Dynamic Catholic*, 182.

the knowledge and habits to live in his love and walk in his ways, and God wants to *send* you out into the world to share his love with others."[111] He adopts this as an approach to evangelization. He notes that "even among highly engaged Catholics, evangelization is the weakest link in the chain. When asked to rate themselves between 1 and 10 for each of the four signs, the 7 percent rated themselves at 6.8 percent or higher for prayer, study, and generosity, but at just 4.9 percent for evangelization."[112] In other words, many Catholics who feel comfortable or reassured in prayer, study, or generosity still struggle in the area of evangelization. By developing the *win, build*, and *send* model recommended by Kelly, I will challenge the participants to see how the foundation established through the first three signs can motivate our evangelization approach. I have chosen this strategy because it presents some simple steps that can help active Catholics navigate the work of evangelization. The process will facilitate teamwork (weekly prayer and study together) and training (gift discernment that creates an awareness of each person's gift and evangelization).

As part of my strategy, I will adopt a ten-minute rule to facilitate a growth process in each sign. The ten-minute rule will require prayer for at least ten minutes each day and Bible study for ten minutes each day. The ten-minute rule will

---

[111] Ibid., 162.

[112] Ibid., 187.

also help participants strive to be generous by sharing their gift for at least ten minutes and reaching out to share the faith with an inactive Catholic or a non-committed person at least once each week. The ten-minute rule for engaging an inactive Catholic or the unaffiliated will enable members to build a body with the person to arrive at a place where they can offer an invitation to participate in our weekly liturgy. As regards prayer, participants will start with a short vocal prayer, a two-minute scripture reading, and a six to seven-minute time of silence or quiet meditation. Doing this will encourage a daily routine in the life of participants. However, the goal is to challenge members to gradually build a habit and progressively engage more in prayer for their spiritual development.

3.5: Sherry Weddell's Contribution to Lay Evangelization

Sherry Weddell is the co-founder and executive director of the Catherine of Siena Institute. Her intentional discipleship program focuses on prayer and discernment. She has organized and trained many Catholics, clergy, and laity in her "*Called and Gifted* workshop."[113] She notes in her writ-

---

[113] Sherry Weddell, *Becoming A Parish of Intentional Disciples* (Huntington, IN: Our Sunday Visitor Publishing, 2015), 37.

ing the striking connection between Faith and Fruits.[114] In a brief introduction to the Institute's website, it states:

> The Catherine of Siena Institute (CSI) is an affiliated ministry of the Western Dominican Province dedicated to equipping parishes for the evangelization and formation of lay Catholics for the sake of their mission to the world. The Institute provides innovative resources and programming that are faithful to the teaching of the Church and will enable parishes to become lively and dynamic centers of evangelization and formation. Since our founding in 1997, we have worked with over 100,000 lay, religious, and ordained Catholics in hundreds of parishes in over 150 (arch)dioceses on five continents. The Church clearly teaches that all of the baptized are called by Christ to proclaim his Gospel in the world.[115]

---

[114] Sherry Weddell, *Fruitful Discipleship: Living the Mission of Jesus in the Church and the World* (Huntington, IN: Our Sunday Visitor Publishing, 2017), 39.

[115] Sherry Weddell, *The Catherine of Siena Institute* Web post, September 10, 2020 [2:40 p.m.], accessed September 10, 2020, https://siena.org/fid-formation-days

Weddell's work on discipleship is at the core of her evangelization program. She describes discipleship as a "lived spiritual experience."[116] Weddell further describes this lived spiritual experience in the context of our daily encounters when she writes that

> Disciples pray with passion. Disciples worship. Disciples love the Church and serve her with energy and joy. Disciples give lavishly. Disciples hunger to learn more about their faith. Disciples fill every formation class in a parish or diocese. Disciples manifest charisms and discern vocations. They clamor to discern God's call because they long to live it. Disciples evangelize because they have really good news to share. Disciples share their faith with their children. Disciples care about the poor and about issues of justice. Disciples take risks for the Kingdom of God.[117]

In other words, discipleship is the process of engagement in one's Catholic faith and desire to share the faith with other Catholics or people of no faith affiliation, either in one's family or outside one's circle of friends. Weddell notes that "Father Michael Sweeney, O.P., and I founded

---

[116] Weddell, *Forming Intentional Disciples*, 11.

[117] Ibid., 81.

the Catherine of Siena Institute, a ministry of the Western Dominican Province. Our mission is to equip parishes to form lay apostles. Our working assumption at the time was that most Catholics just needed solid apostolic formation to discern and answer God's call. We used the *Called and Gifted* discernment process I had developed [in 1993] as a way to introduce Catholics to the idea that they were apostles with a mission by virtue of their Baptism."[118] Her work with Catholics continues to be a means for training the clergy and laity in discerning their gifts to contribute to the mission of evangelization.

3.6: Weddell's Called and Gifted Discernment Process

Since her work on discernment helps Catholics recognize their gifts and support the Church's evangelization mission, Kelly's approach enables me to develop a process that will help participants arrive at the point of discernment. Sherry Weddell's *Called and Gifted* program assists the laity and clergy in discerning their charisms for the Church's mission through the following three-step process: Foundation, Interview, and Discernment. At the end of the three steps in the *Called and Gifted* Discernment process (Foundation, Interview, and Discernment), the participants attend a Team Training workshop to enhance their work-

---

[118] Ibid., 9.

ing as a team to engage inactive Catholics in the parish community and reach out to non-Catholics. The entire process allows participants to test a single gift or charism as they work with others with similar gifts as a team. The discernment of each person's gift within a team enhances confidence in individuals and facilitates the engagement of the gift for the growth of the parish community.

Moreover, the entire process with the various options that flow from it trains Catholics to be disciples and appreciate the process of discernment. Participants from Weddell's work on intentional discipleship note that "we have learned about evangelization and making disciples from working with 130,000 Catholics in over 500 parishes in what we call FID (Forming Intentional Disciples) Formation Days. Learning included:

- How our broader culture affects evangelization
- How to recognize the stages of another's spiritual journey;
- How to have spiritual conversations with people of all religious backgrounds or none;
- The first hopeful signs of change in a parish; and

- The practical first steps parish leaders take when setting out to make their communities a center of discipleship."[119]

In thinking about how the culture affects Catholics today, Weddell shares how the process of discernment creates an opportunity to listen to Catholics sharing their faith and, for many Catholics in the program, their first time talking to someone about their faith. In her book, *Forming Intentional Disciples*, Weddell notes:

> As we listened to the spiritual experiences of tens of thousands of Catholics, we began to grasp that many, if not a majority of Catholics, don't know what "normal" Christianity looks like. I believe that one reason for this is the selective silence about the call to discipleship that pervades many parishes. Catholics have come to regard it as normal and deeply Catholic to *not* talk about the first journey—their relationship with God—except in confession or spiritual direction. This attitude is so perva-

[119] Sherry Weddell, *The Catherine of Siena Institute* Web post, September 10, 2020 [2:45 p.m.], accessed September 10, 2020, https://siena.org/fid-formation-days.

> sive in Catholic communities that we have started to call it the culture of "Don't Ask, Don't Tell."[120]

The context within our churches and the general perception of silence about our faith challenges Catholics. For many Catholics, sharing their faith or acknowledging that they are Catholics is sometimes an uncomfortable experience.

Nevertheless, unless one is courageous to share their faith even with another Catholic, discerning the depth of one's faith can be difficult. Weddell adds that:

> Widespread neglect of the interior journey of discipleship has unintentionally fostered an immense chasm between what the Church teaches is normal and what many Catholics in the pews have learned to regard as normal. Many lifelong Catholics have never seen personal discipleship lived overtly or talked about in an explicit manner in their family or parish. It is difficult to believe in and live something you have never heard anyone else talk about or seen anyone else live.[121]

---

[120] Sherry Weddell, *Forming Intentional Disciples*, 56.

[121] Ibid., 57.

If we grow by speaking and living our faith, it is worthwhile to encourage Catholics to do so even among themselves as a starting point. The more we live out what we believe as Catholics and gradually become comfortable in doing so, it will become easier to share with other Catholics and a step towards evangelization. Weddell further explains that for Catholics to come to the point of having a spiritual conversation with either another Catholic or someone unaffiliated, a process of bonding needs to take place over a period leading to some trust. She notes:

> Once a bridge of trust is in place, people can be surprisingly open to what we have termed "threshold conversations." What do we mean by threshold conversation? It is the initial step of *listening* evangelism, that moment when we invite an individual to talk, simply and directly, about his or her lived relationship with God. We listen prayerfully, seeking to learn what the journey has been like and, if possible, what she or he is at now. If we take the time to first learn where people have been and where they are now, we can much more effectively encourage them on the journey to intentional discipleship.[122]

---

[122] Ibid., 190.

Accordingly, when conversations begin to occur and trust grows, there will be signs of engagement within the parish community. As engagement increases, active Catholics will feel more comfortable reaching out to inactive Catholics, and gradually a process of trust and conversation will lead to outreach and evangelization. These conversations that can be informal in appearance can intentionally engage the parish community in its evangelizing effort.

I will explore some of her work on discernment in my strategy to enable participants to recognize some of the gifts they possess. Doing this will challenge participants to be generous with their charisms in preparation for evangelization. I have not seen any documents that highlight Church growth through her program of forming intentional disciples. Still, this program's benefit is crucial since it reminds every disciple of an essential aspect of our faith—the charisms we receive as members of the Body of Christ. The theology of Charisms from St. Paul's writings and the Vatican II documents on the laity's work and ministry guide her work. She frequently refers to John Paul II's writing, *Christifideles laici,* on the vocation and mission of the lay faithful in the Church and the World.[123]

The mission of evangelization requires some degree of self-knowledge regarding one's faith. Sherry Weddell's three-step discernment process facilitates proper self-

---

[123] Weddell, *The Catherine Siena Institute,* Web post.

knowledge that prepares an individual for the work of evangelization. This process includes setting up the foundation for the discernment of gifts, an interview process that helps each person think through their particular gifts and discern their specific gifts. She writes that "part of the art of discernment is learning to listen to the urge to do what is good, moment by moment, to cultivate an open and responsive ear to God's voice."[124] Weddell makes a case for the importance of listening to God's voice through the active process of discernment. As part of my strategy, I will incorporate into our intervention process for this project the following steps: "examining one's inner experience through prayer, evaluating the effectiveness of our strategy of prayer and study, and weighing the responses of fellow trainees."[125]

While Weddell's work is essential, I will utilize her approach to discernment in my strategy on generosity as it concerns my intervention for the sake of my context. I have chosen this because her work facilitates the exercise of each person's gift as it affects evangelization. The discernment process provides the laity an opportunity to recognize that with the gifts each person receives either through baptism

---

[124] Sherry Weddell, *Discerning Charisms: A Workbook for Navigating the Discernment Process* (Colorado Springs, CO: The Siena Institute Press, 2002), 5.

[125] Weddell, 17.

or confirmation, we are stewards and not authors of the gifts of God.

In addition, her discernment program facilitates team-building among members who possess similar charisms. The team-building process in her work will enable members of the laity to develop bonds that promote more effortless engagement of inactive Catholics and the unaffiliated. The team-building process will be essential in creating an atmosphere where members can support each other and engage inactive members or non-Catholics who may show interest in either returning to the faith practice or desire to know more about the Catholic faith.

Nevertheless, her contribution to evangelization promotes the laity's evangelization through the sharing of their gifts. Recognizing each person's gift facilitates a division of labor that allows for total participation and involvement of members by sharing their gifts. Therefore, the discernment process is both renewing and engaging members who need a reminder of the gifts they possess through the reception of the Sacraments.

3.7: Conclusion

Since my thesis project will engage both Kelly and Weddell, their resources will help members become dynamic in their faith and intentional in their discipleship. These resources from Kelly and Weddell's work will be advantageous to our rural parish community, considering that the Catholic population is a small percentage of the entire population. When members become more confident in their faith, they will be more willing to share their faith with the inactive Catholics and the unaffiliated.

The next chapter of this thesis will focus on the effective use of social media in the Church's mission of evangelization. I will highlight how the Second Vatican Council, in its Decree on Social Communications, *Inter mirifica*, and contemporary Popes, promoted the use of social media in the Church's mission to spread the gospel in the Church and world. I will also discuss some leading Catholic social media evangelists whose resources and writings contribute to my ministerial intervention that uses social media.

Chapter 4

Social Media and Evangelization

4.1: Introduction

In the third chapter, my primary focus was on Matthew Kelly and Sherry Weddell, the leading lay evangelists whose work and resources are the foundation of my ministerial intervention. Also, in chapter 3, I briefly described the lay evangelists whose writings and works enrich the discussion on lay participation in evangelization. As for Kelly and Weddell, I gave a preview of the four signs of a dynamic Catholic and the Called and Gifted discernment process, respectively, critical for my ministerial interventions.

In this chapter on social media and evangelization, I will begin my discussion by looking at how the encyclical letters by Pius XI, *Vigilanti cura*, and Pius XII's *Miranda porsus* set the stage for the discussions in Vatican II on the Decree on Social Communications, *Inter mirifica*. After discussing the ways Vatican II promotes the use of social media for the mission of the Church, I will briefly highlight some recent papal statements from the World Day of Communication that challenge Catholics to engage in social media and how these engagements can enhance our evangelization. These statements are crucial in understanding the continuity from Vatican II and the push to see how the Church through its members can incorporate social media as a tool for evangelization. Next, I will discuss some significant insight into social media for evangelization coming from Catholic evangelists, including Brandon Vogt, Clarissa Aljentera, Eugene Gan, Shaun McAfee, and Bishop Robert Barron. From the midst of these evangelists, I will highlight a leading voice in the discussion on using social media, Meredith Gould, and show her insights contribute to my ministerial intervention. I will also elaborate on the concept of networking and how it can benefit the Church's mission. At the end of these discussions, I will outline the implications of social media for my ministerial context and offer a brief conclusion.

4.2: Early Papal Encyclicals & Vatican II on Social Media

The advent of social media and its benefits to evangelization is undeniable when looking at the number of daily social media users. Besides its use, it provides a unique opportunity to engage the secular culture and be a presence to a diversity of people who may be open to receiving our invitation to encounter Jesus Christ in the Catholic Church.

During the pontificate of Pope Pius XI (6 February 1922 – 10 February 1939), the influence of the media was notable. It was an opportunity to invite the Church, especially its leadership, to be mindful of the ongoing developments. In his encyclical *Vigilanti cura*, Pope Pius XI notes that

"Here is a matter for which immediate provision is absolutely necessary: we must ensure that all progress made, by God's favor, both in human knowledge and in technical skill, shall in practice so serve God's glory, the salvation of souls and the extension of Christ's kingdom, that we all, as the Church bids us pray, may so pass through temporal goods that we may not lose what is eternal."[126] So, even back in the early twentieth century, the Church recognizes that media is essential for evangelization.

A couple of decades later, Pope Pius XII (2 March 1939 – 9 October 1958) also acknowledged the impact of the media and how the Church can benefit from its use. In his encyclical, *Miranda porsus*, Pius XII notes that these means of communication can "be at the service of truth in such a way that the bonds between peoples may become yet closer."[127] In expressing this invitation, the Holy Father highlights how social media tools can be a source of proclaiming the truth and evangelizing the culture.

---

[126] Pius XI, Encyclical Letter on the Motion Picture, *Vigilanti cura* (29 June, 1936), Web post accessed online 22 July, 2021 at https://www.vatican.va/content/pius-xi/en/encyclicals/documents/hf_p-xi_enc_29061936_vigilanti-cura.html

[127] Pius XII, Encyclical Letter on the Motion Picture, Radio and Television, *Miranda porsus* (8 September 1957), Web post accessed online 22 July, 2021 at https://www.vatican.va/pius-xii/documents/hf_p-xii_enc_08091957_miranda-porsus.html

While focusing on the media of the times—film, the press, radio, and television—these encyclicals provide some foundation for the discussions at Vatican II on social media. As one thinks of the conversations that arise from social media use within and outside the Church, looking at the speed of these developments in our contemporary society and how The Council invites the laity's involvement is essential. Vatican II's Decree on the Media of Social Communication, *Inter mirifica,* acknowledges the variety of social media communications "contribute greatly to relaxation, the enrichment of people's minds and the spread and consolidation of the kingdom of God."[128] The Council, in the same discussion, invites the pastors to involve the laity and, when possible, train and instruct the laity on the proper means of engaging social media as a tool for evangelization. *Inter mirifica* notes that "its task involves using the media to proclaim the good news of salvation as well as teaching people how to use them properly. The Church has an innate right to use and own any of these media, which are necessary or valuable for the formation of Christians and for all of its pastoral work. It is the duty of pastors of

---

[128] Vatican Council II, Decree on Social Media, *Inter mirifica* (4 December 1963) 2, in *Vatican Council II Constitutions and Declarations: A Completely Revised Translation in Inclusive Language,* edited by Austin Flannery (Northport NY: Costello Publishing Co., 1996).

souls to instruct and direct the faithful in their use, so that they will contribute to their own salvation and perfection and that of all of humanity.[129] The directive invites pastors and those in leadership positions within the church to use all available opportunities to engage other Catholics on the importance of social media in Catholic evangelization.

The Council gives us a foundation for engagement. It proposes how social media engagements can provide an opportunity for Catholics—active and inactive—to grow in their appreciation of the faith and the sacraments. The Decree also encourages all the faithful to acquire the knowledge and skills to make this mission possible. The Decree further highlights that "All the members of the Church should make a concerted effort to ensure that the media are utilized in the service of the many works of the apostolate without delay and as energetically as possible, where and when they are needed."[130] The urgency in which the Decree invites all—clergy and laity—further emphasizes the importance of social media in the Church's mission. The Council statements also indicate how the developments in social media provide an opportunity for collaboration between the clergy and the laity regarding evangelization. The Decree continues by noting that "Lay people who work professionally in the media should endeavor to

---

[129] IM 3.

[130] IM 13.

bear witness to Christ: first of all, by doing their work competently and in an apostolic spirit, secondly by collaborating directly, each one according to her or his ability, in the pastoral activity of the Church, making a technical, economic, cultural, or artistic contribution."[131] The collaborative effort between the laity and clergy, especially at this stage in the Church's history, is significant when one considers the involvement of the laity due to previous institutional challenges within the Church.

Moreover, the laity has an edge in the development and engagement in this area since many may likely have professions that include work in social media. Collaboration with members of the laity who are well-versed in social media and its operation is necessary throughout this process and an opportunity to share ideas that will enhance evangelization and the Church's mission. *Inter mirifica* goes on to note that "Since people of different ages and cultural backgrounds have access to the media, they need theoretical and practical formation in their proper use, a formation tailored not merely to the character of each medium but to the needs of each group."[132] While the pastors may know the theological foundation that guides the use of social media, the laity's involvement and collaborative presence will enhance the engagement of the target audience for our evangelization. Like

---

[131] Ibid.

[132] IM 16.

the decree notes, the training process needs to be in a format that addresses the needs of each community.

4.3: Promoting Social Media through World Communications Day

The early encyclical letters and the Decree from Vatican II all express the impact of the media in our engagement and the opportunity it provides for evangelization. The establishment of World Day for Communication provides an opportunity for all—clergy and laity—to focus on recent developments while gauging how the Church can respond to these changes.

Consequently, social media enlightens the clergy and the laity on specific areas to aid evangelization and facilitate the Church's mission. Pope Paul VI, in his message for the 8th World Communications Day given on 16 May 1974, notes that "Evangelization is an integral part of the mission of the Church, sent by Christ into the world to preach the Gospel to every creature (Mk 16:15). The Church carries out this duty above all in her liturgical life, but she is constrained to fulfill it also in all the ways and by all the means which are available to her to use as she abides among the people of each continent."[133] Pope Paul VI expresses that

---

[133] Paul VI, 8th World Communications Day, *Social Communication and Evangelization in Today's World* (16 May 1974),

while the liturgy is at the heart of our life as Catholics, our engagement will require using means outside the liturgical context to evangelize.

The establishment of the Pontifical Council for Social Communications in September of 1948 under the title; Pontifical Commission for Educational and Religious Films provided the Church a means to review and engage the developments in the media. In his Decree to the Roman Curia establishing the Commission, Pope Pius XII notes that "The Pontifical Commission for Educational and Religious Films examined the complex educational and pastoral problems of the audiovisual era just coming into being and quickly realized that action was going to be needed."[134] The Council was to engage the media, which includes reviewing films, news, and television broadcast. The name of The Council was changed during Vatican II on 2 April 1964 from Pontifical Commission for Educational and Religious Films to the Pontifical Council for Social Communications.

Moreover, up until its dissolution and merging with the Secretariat for Communication to create the Dicastery for

---

Web post accessed 22 July, 2021 at https://www.vatican.va/ content/paul-vi/en/messages/communications/documents/hf_p-vi_mes_19740516_viii-com-day.html

[134] A Brief Outline and History of the Pontifical Council for Social Communications, Web post accessed 13 August 2021 at https://www.vatican.va/roman_curia/pontifical_councils/pccs/documents/rc_ pc_pccs_pro_14101999_en.html

Communications in March 2016, the Pontifical Council for Social Communication was essential in offering updates for the Church's progress in social media engagement. Almost thirty years after Vatican II, the Pontifical Council, in its pastoral instruction, *Aetatis novae,* notes that "At the dawn of a new era, a vast expansion of human communications is profoundly influencing culture everywhere. Revolutionary technological changes are only part of what is happening. Nowhere today are people untouched by the impact of media upon religious and moral attitudes, political and social systems, and education."[135] Even after forty years of engagement, the impact of social media is still essential for the Church's mission and evangelization.

In his message for the 36th World Communications Day, Pope John Paul II reminds all the faithful that "The Internet causes billions of images to appear on millions of computer monitors around the planet. From this galaxy of sight and sound will the face of Christ emerge, and the voice of Christ be heard? For it is only when his face is seen, and his voice heard that the world will know the glad tidings of our redemption. This is the purpose of evangeliza-

---

[135] Pontifical Council for Social Communication, Pastoral Instructions on Social Communications, *Aetatis novae* (22 February 1992) 1, Web post accessed online at http://www.vatican.va/roman_curia/pontifical_councils/pccs/documents/rc_pc_pccs_doc_22021992_aetatis_en.html

tion."[136] The internet has become a thriving platform for every kind of communication. As evangelizers, its use invites us to see the opportunity as one that facilitates reaching a unique audience. It would be difficult not to notice the far-reaching possibilities this has on a parish community.

In recent times, Pope Francis continues to encourage the clergy, religious, and the laity to engage in social media communications and adopt communication as a means to connect with all peoples. In his message for the 50th World Communications Day, he notes that "Communication has the power to build bridges, to enable encounter and inclusion, and thus to enrich society. How beautiful it is when people select their words and actions with care, in the effort to avoid misunderstandings, to heal wounded memories and to build peace and harmony. Words can build bridges between individuals and within families, social groups and peoples. This is possible both in the material world and the digital world."[137] Pope Francis sees social media as an essen-

---

[136] John Paul II, 36th World Communications Day, *Internet: A New Forum for Proclaiming the Gospel* (12 May 2002), Web post accessed online August 6, 2019 at http//w2.vatican.va/ content/john-paul-ii/en/messages/communications/documents/hf_jp-ii_mes_20000124_world-communications-day.html.

[137] Pope Francis, 50th World Communications Day, *Communication and Mercy: A Fruitful Encounter* (24 January 2016), Web post accessed online 6 August 2019 at http:// vatican.va/con-

tial pathway to peace and harmony between peoples by highlighting this point.

Therefore, part of the service we offer as disciples is reaching out through social media to provide those seekers with the Good News while being present in cyberspace or the virtual community. Pope Francis describes this kind of engagement as "forming a culture of encounter."[138] The messages and exhortations from the Pontifical Council and the Popes remind all the people of God of our contemporary context and how our engagement in social media can promote the Church's mission.

4.4: Contemporary Sources on Social Media

In this section, I will highlight some of the most prominent social media evangelists in the Catholic Church today. In engaging social media, these leading Catholic voices include; Shaun McAfee, Eugene Gan, Clarissa Valbuena Aljentera, Brandon Vogt, Bishop Robert Barron, and Meredith Gould. Their contributions have enhanced my ministerial intervention and decision to apply social media as a

---

tent/francesco/en/messages/communications/documents/papa_francesco_20160124_messaggio-communicazioni-sociali.html

[138] John C. Cavadini and Donald Wallenfang, eds., *Pope Francis and the Event of Encounter* (Eugene, OR: Pickwick Publications, 2018), 252.

tool for evangelization. Particularly, I will discuss Meredith Gould extensively and how her contributions present the most helpful resource for my rural parish community engagement.

Shaun McAfee is a convert to Catholicism and a devoted blogger. Besides his interest in Catholic evangelization and dedication to apologetics, he is the founder of EpicPew.com. He shares his writings, videos, and reviews regarding issues that can strengthen the faith of Catholics on his weekly blog.[139]

In his book *Social Media Magisterium*, McAfee states, "Communication is the chief purpose of the media in its various roles of informing, educating, training, or entertaining."[140] While the media can serve multiple purposes, as McAfee notes, it is essential to see its engagement as one that can benefit a parish or a diocese in its evangelizing mission. Mindful of these roles of the media, McAfee adds that "By learning how best we should communicate corporately as a Church or individually as representatives of the Body of Christ online, we are that much better prepared to disciple others, preach the gospel, and be attractive stew-

---

[139] Shaun McAfee, *Sophia Institute Press*, Web post accessed 23 July 2021 at https://www.sophiainstitute.com/author/shaun-a-mcafee

[140] Shaun McAfee, *Social Media Magisterium* (St. Louis, MO: En Route Books, 2018), 22.

ards of the Kingdom."[141] Applying social media to evangelize can promote the Church's mission, especially in the parish context.

Moreover, if parishes desire to adopt social media as a means of engagement, McAfee gives a few valuable recommendations through his years of blogging and online engagement. He notes that "I always tell them that any tool can be misused. A tool is a tool, and what makes the tool useful is the user. The user must use the tool in three ways: according to its purpose, with intention, and with skill. Catholics may use the media rightly and successfully if they employ these three rules. The media has a sure purpose, and that's communication with others."[142] The proper use of social media, especially when we engage—either Catholics or the unaffiliated—makes a difference.

Eugene Gan is another contemporary Catholic contributor who supports social media tools and is presently an associate professor of interactive media communications at the Franciscan University of Steubenville. [143] Besides his articles for the Pontifical Council for Social Communications, he is an expert on many aspects of new media technology.

---

[141] Ibid.

[142] Ibid., 179.

[143] Eugene Gan, *Infinite Bandwidth: Encountering Christ in the Media* (Steubenville, OH: Emmaus Road Publishing, 2010).

In his writing, Gan notes that "If human beings were fish, media would be the sea in which we swim. It's everywhere, shaping how we work, how we learn, how we communicate, how we entertain ourselves, and even how we think."[144] In essence, the effect of social media on our lives goes beyond the fact that we are Christians but our social nature as humans. Gan also echoes the words of McAfee when he notes that "media and media technology are unquestionably everywhere. But that's not necessarily a bad thing. When used rightly, media can do an untold amount of good."[145] An intentional engagement of social media as a tool for evangelization can affect our audience in ways that may strengthen their faith or increase a desire to know more about the Catholic faith. Gan, therefore, notes that "For very good reason, the Catholic Church has called media technologies' gifts from God.' They are tools that can and have been used to connect us, enlighten us, and transform us."[146] The connection that comes from social media, especially with an increase in its use in our contemporary society, makes it an essential means of evangelization.

Clarissa Valbuena Aljentera is a contributor to America Magazine and Our Sunday Visitor. Through her writings and seminars, she encourages the laity to integrate social

[144] Ibid., 8.

[145] Ibid.

[146] Ibid., 9.

media into their conversations to facilitate a desire to evangelize.[147] She has a passion and energy for communication and enjoys sharing with others.[148]

In her writing, Aljentera emphasizes the importance of social media to all facets of parish ministry. Commenting on how parish ministry can evolve and offer new ways of engagement, Aljentera, in her book, *The Parish Guide to Social Media*, writes:

> With social networks, communication through media is no longer one-sided. In the past, people who watched television and listened to the radio were not able to interact with each other. Technology has made it possible for people to respond in immediate fashion to something they read, watch, or listen to. In doing so, they become part of the conversation. Through our responses, others get a sense of who we are and what we believe. They begin to understand what is important to us.

Aljentera notes how social media opens up opportunities for conversations and an outreach that informs and en-

---

[147] Clarissa Valbuena Aljentera, *The Parish Guide to Social Media* (New London, CT: Twenty-Third Publications, 2013), back cover.

[148] Ibid.

gages. Through this kind of engagement, the Church has a platform to evangelize and connect with persons who have become inactive or are unaffiliated. Aljentera further notes that "These kinds of faith connections and activities are crucial to the life of your parish and its people, and social media is not a substitute for any of them. But what social media can do is help enhance those faith connections and help to bring more and more people into life-giving relationships within your parish."[149] The relationships that a parish builds through its online presence can further the Church's mission and create opportunities for evangelization.

Brandon Vogt is the founder of ClaritasU, which trains Catholics to talk about their faith, especially hot button issues, and is the Senior Content Director for Bishop Robert Barron's Word on Fire Catholic Ministries.[150] He is a speaker and writer who blogs about theology, technology, social justice, and books at ThinVeil.net.[151]

Vogt, in his writing, maps out a progression in the development of media—old and new—beginning in the scrip-

---

[149] Ibid., 11.

[150] Brandon Vogt, *Brandon Vogt: About*, Web post accessed 23 July 2021 at https://brandonvogt.com/about.

[151] Brandon Vogt, *The Church and the New Media: Blogging Converts, Online Activists, and Bishops Who Tweet* (Huntington, IN: Our Sunday Visitor Publishing, 2011), back cover.

tures and describing the opportunity social media creates for evangelization today. In his book *The Church and the New Media*, Vogt notes that "Christianity reveals a God who uses all sorts of media to get his message across, from pillars of fire to bright stars, from burning bushes to a talking donkey. He delivered the Ten Commandments on chiseled rock, painted rainbows to signal his covenant, and used a cross to display his love."[152] Vogt describes how the crucial moments in salvation history indicate God's communication with humanity, especially his chosen people. Yet, in our contemporary context, Vogt maintains that God still communicates with us through different forms. He notes that "the Bible doesn't mention God using emails, blog posts, podcasts, or e-books – and when the prophet Habakkuk [2:2] quotes God saying, 'write the vision; make it plain upon tablets,' he wasn't promoting iPads, either! Even though these technologies don't show up in Scripture, history has shown that God showers down new mediums at specific times to spread his Word in new ways."[153] Throughout history, the evolution of social media is such that the Church needs to be aware of the various changes and take proactive steps to enhance its engagement in society.

---

[152] Ibid., 15.

[153] Ibid.

Therefore, like other leading contributors, Vogt invites all Catholics to be disciples of social media to engage contemporary society. He concludes by noting that "if the Church's promotion of evangelization, formation, community, and the common good is to continue throughout future generations, she must harness these technologies and utilize them well."[154] The future of the Church's mission will depend on our response to the ongoing changes in social media and embracing the opportunity it offers for evangelization.

Bishop Robert Barron is the founder of Word on Fire Ministries, and besides posting YouTube videos regularly, he maintains an interactive website where he features his writings and sermons.[155] He is presently an auxiliary Bishop of the Archdiocese of Los Angeles and the creator and host of a documentary series on *Catholicism*.[156]

As an ardent social media user, Bishop Barron sees the internet as a unique space for engaging the culture and offering opportunities to persons who have questions to have a reasonable conversation. He states in his book *Arguing Religion: A Bishop Speaks at Facebook and Google,* that

---

[154] Ibid., 21.

[155] Ibid., 28.

[156] Robert Barron, *Arguing Religion: A Bishop Speaks at Facebook and Google* (Park Ridge, IL: Word on Fire, 2018), back cover.

"what has become abundantly clear to me over the years is that, though there is a great deal of energy around matters religious, and though lots of sharp words are regularly exchanged, very few of the people on the internet really know how to have an argument about religion."[157] While social media avenues present us with a space for interaction, it also requires a different approach to engagement and conversations that deal with religion. Bishop Barron, in his writing, also challenges Catholics who desire to engage in this new form of media to recognize that there are benefits to engaging in social media. He writes:

> I've been engaged in the work of evangelizing the culture. I've written ten books of theology [at the time] and spirituality; I've taught courses in philosophy and systematic theology at Mundelein Seminary, one of the largest seminaries in the United States; I've broadcast and podcasted over 500 sermons; I've lectured business and civic leaders, and I've published dozens of articles in learned and popular journals. But I believe that the most effective work I've done in this arena is through the internet.[158]

---

[157] Ibid., 4.

[158] Vogt, *The Church and the New Media*, 27.

From Bishop Barron's standpoint, the internet has been his most effective means of engagement throughout his years of evangelization. Since establishing his *Word on Fire Ministry,* Bishop Barron has used social media to engage the culture and reach out to Catholics, Protestants, the unaffiliated, agnostics, and people who describe themselves as atheists. He considers the opportunity that social media platforms offer essential for evangelization and engaging people who may never have a chance to encounter the Church or its representative. The context may be different, as he notes, but the benefits of using social media, in the long run, will be worth the challenge.

4.5: Meredith Gould's Contributions to My Ministerial Intervention

Meredith Gould is a Jewish convert to Catholicism, a digital strategist, sociologist, and digital communications professional. She is nationally known as a passionate advocate of using digital tools for ministry and encourages practical spirituality.[159] Gould highlights the importance of social media as a means for evangelization within the Church.

---

[159] Meredith Gould, *The Social Media Gospel: Sharing the Good News in New Ways* (Collegeville, MN: Liturgical Press, 2013), back cover.

From her experience as a convert, she shares how her spiritual journey contributes to an ongoing desire to connect with those searching for spirituality. Her mission to connect and share her stories is made easy through the presence of social media. In her book *Desperately Seeking Spirituality*, Gould notes that "Depending on who you ask I was either 'always' or 'never' like this and by 'like this,' I mean spiritual. There were some hints along the way during childhood and early adulthood, but my impetus to seek something greater than and beyond self kicked into gear during my late thirties."[160] Her spiritual journey and conversion experience, as she shares, has become the opening that continues to inspire her social media engagement.

In her book *The Social Media Gospel*, she emphasizes that social media is vital for proclaiming the Gospel, hence, the title of her book. Gould writes that "Social media are web-based tools for interaction that, in addition to conversation, allow users to share content such as photos, videos, and links to resources."[161] She sees social media as a means of transcending generations and depending on the platform, and it can be highly engaging and effective. She writes that "using what we know about generations will make it

---

[160] Meredith Gould, *Desperately Seeking Spirituality: A Field Guide to Practice* (Collegeville, MN: Liturgical Press, 2016), xv.

[161] Ibid., 3.

easier to work together within and for [the] church."[162] In her work, she goes through many of the widely used social media applications like Instagram, Twitter, YouTube, Facebook, LinkedIn, Blogging, Pinterest, and Snapchat. For this project, I will use Facebook and rely on Gould's knowledge of social media because it is the most widely used social media platform in my ministerial context. Gould notes, "Christianity is a religion anchored in community. Jesus set the standard by declaring God present when two or three are gathered (Matt 18:20). And I cannot resist pointing out how Jesus was calling us friends long before Facebook' friending' was invented."[163] Building community through social media and educating members of a parish community to see its benefit for evangelization requires following the words of Jesus that invite us to gather.

For effective evangelization, Gould recognizes how social media communication enables us to extend God's presence through the media while building relationships that engage those who have an online presence within and outside the community. To this point, Gould notes:

---

[162] Meredith Gould, *Transcending Generations: A Field Guide to Collaboration in Church* (Collegeville, MN: Liturgical Press, 2017), 22.

[163] Gould, *The Social Media Gospel*, 13.

> Social media tends to connect rather than isolate users. Relationships starting as conversations via social media frequently lead to in-person meetings as geography and time allow. Email exchanges and phone conversations usually, but not always pave the way to meeting IRL (in real life). Do not assume that whatever begins on social media stays on social media. Now that social media has become a normative tool for communication, it's common for individuals who first meet in a physical environment (e.g., church) to "friend" and "follow" each other to continue the conversation via social media.[164]

Social media's relationship and engagement can create bonds beyond the parish church, the parish hall, or even the parish boundaries. Her insight into how social media affects a community can easily be the experience of every parish with an online presence.

Since Facebook allows for visual and audio communication, targeting visual and audio learning forms will enhance listening. By listening, Gould notes in her book *Transcending Generations*, that "I do mean *listening*, which involves seeking to understand input as opposed to *hearing*, merely the passive reception of sound."[165] Her emphasis on listen-

---

[164] Ibid., 32.

[165] Gould, *Transcending Generations*, 73.

ing is significant when thinking of creative ways to engage inactive Catholics and the unaffiliated.

The effective use of social media can also facilitate engaging a wider audience, as much data today proves. Gould, writing on the effects of Facebook, notes that "Simply put, in a grandiose but accurate way, with 1.35 billion monthly active users as of October 2014, Facebook, which also owns Instagram, WhatsApp, and video platform QuickFire Networks, is the largest social networking platform in the known universe."[166] These statistics that Gould offers here reflect the time of the publication of her book in 2013. Besides the opportunity Facebook offers, its use within the Church can effectively facilitate outreach to the unaffiliated and inactive Catholics.

Furthermore, through Gould's knowledge of social media, she recommends a series of background work to challenge parishes in social media engagement. She proposes the following questions as guiding principles when thinking of social media and writes:

> Who do we most want to reach? Remember to get information about how your audience receives and absorbs information. You probably want to reach more than one group of people, so who, in order of importance, are your intended audiences?

[166] Gould, *The Social Media Gospel*, 73.

> What do we want to happen as a result of using these tools? While engagement (I.e., interaction) is evidence of social media success, you'll need to clearly define the actions and behaviors you want to emerge because congregants are engaging with you and one another via social media. Are your goals realistic? Are they actionable?
>
> What do we want people to know or learn about us? Do not squander precious online space to tell visitors that your church is . . . a church. Or a "friendly," "welcoming" one. Develop clear distinctive messages.
>
> How will we know we've succeeded? Remember that although these tools speed up communication and make it possible for groups to form quickly, building community always takes longer than anyone expects. How will you respond if your parish council, vestry, church board, or some other group with authority asks if social media is working? Don't rush this process.[167]

The questions presented by Gould provide parishes and their ministerial teams' ways of thinking about their goal of using social media. Also, the intentional use of social media

---

[167]Ibid., 41.

offers a way of notifying members of formational opportunities for evangelization and development.

Through her work on social media, Gould attests that navigating this field is not easy, but, in the long run, its effectiveness will be worth the effort. She writes that "Facebook is great for Church communication and ministry because, in addition to providing an efficient, cost-effective way to broadcast news and publicize events, it provides many other options for generating and sustaining community (e.g., the timeline, groups, customizable tabs on the home page, notes, lists). Posting announcements and prayer requests to the news feed, clicking 'like,' and adding comments are the most obvious ways to use Facebook.[168]In sustaining a parish community, evangelizing through Facebook may be a reasonable means to engage the community. Nevertheless, it is essential to know how social media can help us achieve active participation and increased membership through evangelization.

4.6: Networking as an Important Part of Evangelization

One of the areas whereby social media impacts the lives of its users is the aspect of networking. While this could certainly have various meanings, for my thesis, I will describe networking within my ministerial context as a way of

---

[168] Ibid., 74.

dialogue, interaction, and connectivity among members within the community, including both Catholics and the unaffiliated. By this, one can observe how diverse cultures and languages can connect and exchange ideas on a specific platform without prior knowledge of each other. It is remarkable how social media continues to affect the way we communicate in our contemporary society. The influence of networking is not lost to modern bloggers and social media experts, as is evident in their writings. Vogt, in observing how networking influences communication, notes that "No great evangelists of the past two millennia could have conceived that within minutes they could have their messages beamed to billions of people across the world, cheaply and instantly. St. Paul, the early Church Fathers, St. Francis Xavier, and Archbishop Fulton Sheen each would have given their right arm for access to our New Media."[169] The notable difference from networking is that social media use gives many people access to a diverse audience. For some early social media proponents, how networking shapes our relationships is not unique but only more visible due to the greater use of these tools today. Aljentera notes that "Social networks have changed how we view communication. We expect to be part of the conversation instead of simply listening or receiving information. Organizations invite us to post comments, thus encouraging dialogue with

[169] Ibid., 199.

a broader audience. We are asked for our thoughts and opinions and then offered a venue for sharing them immediately."[170] Thus, networking implies a two-way engagement that facilitates dialogue. In a context of mutual respect, a social media user can post an idea or thought, hoping that there will be a meaningful response from another participant. The opportunity to network with other social media users can benefit evangelization in today's climate, where users cut across all spheres of life, belief, and generation. Vogt accordingly notes:

> These technologies have birthed a 'springtime' of evangelization in the world. People who would never consider setting foot in a church are dialoguing with priests on YouTube. Streaming videos and interactive websites encouraging inactive Catholics to return to the Church have already produced staggering results. And New Media is connecting the Church with many other difficult-to-reach groups: youth, young adults, the elderly and homebound, and those living in remote locations.[171]

The reality of social network for Catholic evangelizers is that it gives us a context to share the Good News and pre-

---

[170]Aljentera, *The Parish Guide to Social Media*, 10.

[171] Brandon Vogt, *The Church and New Media*, 199.

sent to persons searching for a community of faith. Therefore, it is essential to encourage Catholics to engage in networking to promote evangelization and the Church's mission through social media.

4.7: Implications for Ministerial Intervention

Social media presents many opportunities to develop and engage men and women searching and open to exploring the Catholic faith. When thinking of social media also, it is vital to highlight the presence of a virtual community. The virtual community is a real presence that attracts many non-churchgoers who sometimes see this space as an opportunity to discuss ideas that impact their lives. These ideas sometimes include their religious belief or faith challenges. In recognizing this opportunity for us as Catholics, our presence in the virtual community enables us to engage and share with those who may desire to learn or indicate some openness. Thanks in part to social media, our Catholic Faith has one more platform where we can share our faith with interested persons or persons seeking dialogue and desire a parish home.

Nevertheless, while the flattening of the earth and creating a global village presents us with many opportunities, there needs to be some formation on approaching persons through social media.[172] Social media can be the first step to connecting with people of different backgrounds and opening interpersonal contact for future dialogue.

Social media is an essential channel to help my rural parish community connect with members who have become inactive or unaffiliated and may be interested in the Catholic faith. The vocation of discipleship requires all members to see how the new ways or methods of evangelization that social media offers can enhance the Church's mission. The goal is to reach out to many parishioners—active and inactive—and the unaffiliated within the region. My plan is to develop and implement a workable strategy within the bootheel and will require some creativity to achieve this goal. Facebook offers an opportunity to communicate the Good News through a simple post on the web to reach or engage a virtual community.

Furthermore, our engagement and use of social media allow evangelizers to develop techniques to reach out to remote and rural communities within the bootheel. It also creates an opportunity for the parishioners to share the message of the Good News. Today, with the use of social media as a tool for evangelization in our modern culture and society, we experience a different kind of challenge in

---

[172] Thomas Friedman, *The World is Flat: A Brief History of the Twenty-first Century* (New York, NY: Farrar, Straus and Giroux, 2005), cover page.

our engagement or propagation of the Good News. In their article on *Catholic New Media Use in the United States*, Mark Gray and Mary Gauthier note that "We already know that Catholics are no less or more likely than Americans in general to use new technology to access digital or online media. The Church's biggest challenge is to encourage U.S. Catholics to use these devices to connect with their faith more frequently."[173] Collaborating with the laity as co-workers and using Facebook, developing a practical way of reaching out to the people in rural Southeast Missouri and possibly other rural communities to evangelize gives us a way to test this mode of engagement.

Due to its wide usage within our parish community and region, Facebook gives us a means of engagement and outreach within our community. As disciples called to fulfill our mission of spreading the Gospel message, the reality of ministry and life is that we are responsible for engaging the culture. The rural context need not be left out in our evangelizing efforts but included as much as possible.

Many who live within these communities reside on vast acreages of property, limiting frequent personal and face-to-face engagement. While distance in a rural community

---

[173] Mark M. Gray and Mary L. Gauthier, *Catholic New Media Use in the United States*, accessed online at http://www.usccb.org/about/communications/upload/Catholic_New_Media_Use_in_United_States_2012.pdf, November, 2012.

could be a limiting factor for evangelization, social media offers a new opportunity to engage. In this context, social media has become a viable tool to bridge the gap of communication and socialization, albeit on a different level of interaction.

In my ministerial setting in Pemiscot and New Madrid counties, as observable in other areas, a more significant percentage of young adults, more so than older members, constantly engage each other through social media. A Catholic presence on these web-based tools is an opportunity to enter their world and to be a presence that offers hope and alternatives that would, in some cases, be life-giving in their search for answers. Our efforts in using social media will build our members' faith and the faith of those persons who, through our evangelizing action, desire to grow in their faith. It can also be a means within the parish of presenting to the unaffiliated a channel for dialogue.

As a pastor, social media facilitates a web presence that makes it easy to engage and encounter the community. Like many other parishes within the diocese, our parish has a Facebook page to engage our parishioners and the community. Therefore, my desire to engage and evangelize using Facebook offers our parish community a platform that has a broad audience in Southeast Missouri. Using this platform will facilitate our connection with persons outside our parish boundaries.

In the Bible belt, the challenge for many members of the laity is catechesis and adequate training for evangelization that builds their confidence to evangelize. Social media engagement offers an opportunity for such formation and renewal. The slow pace of a rural community can be a blessing in disguise to draw members into embracing this opportunity to learn and grow in their knowledge of the benefits of social media. However, if in-person attendance in the Church proves to be a challenge, as was the experience during the pandemic, social media offers an alternative for engagement.

In a rural parish context, social media provides information for the parishioners and creates opportunities for dialogue and intentional evangelization of the inactive Catholics and the unaffiliated.

4.8: Conclusion

As the various sections in this chapter highlight, social media is a vital tool for evangelization to support the Church's mission. It provides an opportunity for a new approach to ministry and offers both clergy and laity a way of engaging inactive members and the unaffiliated. The diversity of social media tools and how they impact different generations can also positively motivate members to learn and engage in their use.

Within my context in Southeast Missouri, forming members on embracing and utilizing social media is a priority. Since the Catholic Church in this region struggles with a decrease in priestly vocations and a lower Church attendance, like many other regions, we need to be creative in our approach.

In conclusion, engaging members within the parish community and evangelizing the inactive and unaffiliated through social media is essential. The ongoing changes and the new trends in the use of social media by people of all ages in our society require that as people of faith, we develop ways of engaging these changes.

The next chapter will offer the outline for my ministerial intervention that includes a nine-week training program. I will show through a series of charts how engagement in the program grew or diminished. I will also highlight the steps taken and the various participants in the program. The next chapter will also offer insights into developing a training program and engaging members through social media to prepare the laity to evangelize. As Gould notes in her book, *The Social Media Gospel*, that

> Christ has no online presence but yours,
> No blog, no Facebook page but yours,
> Yours are the tweets through which love touches this world,

> Yours are the posts which the Gospel is
> shared,
> Yours are the updates through which hope is
> revealed.
> Christ has no online presence but yours,
> No blog, no Facebook page but yours.[174]

If Christ has no online presence but ours, there needs to be no hesitation on our part to engage and evangelize our brothers and sisters. The social media platform we offer online will be a means of making Christ present to others—either inactive members or the unaffiliated.

---

[174] Gould, *The Social Media Gospel*, 9.

Chapter 5

Methodology and Intervention

5.1: Introduction

In the previous chapter on social media and evangelization, I briefly discussed how the encyclical letters of Pius XI, *Vigilanti cura*, and Pius XII's *Miranda prorsus* set the stage for the discussions at Vatican Council II on the Decree on Social Communications, *Inter mirifica*. Also, I reviewed a few statements from the World Communication's Day that challenge Catholics to engage in social media use and discern how these engagements can enhance evangelization.

Furthermore, I discussed some contemporary sources in the Catholic Church that offer insights into the use of social media for evangelization and the benefits involved in its use. I concluded by noting the benefits of networking and the implications of the contemporary contributions of social media on my ministerial interventions and their impact on my project. The discussions from the preceding chapters will therefore be foundational for my ministerial intervention and the development of this project.

In this chapter, I will briefly describe my ministerial intervention strategy and the participants who volunteered for this project. I will further describe my project's weekly process and how the four signs of a dynamic Catholic—

prayer, study, generosity, and evangelization—suggested by Matthew Kelly, are a framework for the program's first three weeks. I will review the weekly sessions and highlight the materials that I shared with participants during these sessions. Included in the weekly sessions, I will discuss how Sherry Weddell's *Called and Gifted* discernment process enhanced the participation of members in the program. Given the discernment by participants, I will briefly describe some of the charisms and how participants were assigned ministerial teams based on these charisms. I will conclude this chapter by indicating the data collection process throughout my research project and a brief conclusion.

5.3: Ministerial Intervention Strategy

To promote parishioner's awareness of their responsibility to evangelize, I offered a nine-week program of formation in evangelization. The program was designed specifically for members who signed up for it. In my first chapter, I observed that a few active members in my parish made a commitment to find ways to evangelize. These members also desired to learn how to engage inactive members and the unaffiliated. Many of them, however, expressed a lack of training and skills to facilitate their commitment to evangelization. My nine-week program design was to offer the participants some basic skills on evangelization based on my research and experience of evangelization. The program will engage some skills and resources tested by other lay evangelists, albeit in different contexts. Still, it would be a simple way to engage in a rural parish community.

In developing this program, my goal was to share some basic principles that will give members who have never engaged in evangelization the foundation to build on as they develop new skills. After extensive research on lay evangelists, I focused my program on two lay experts in evangelization: Matthew Kelly and Sherry Weddell. I chose these lay experts because of the adaptability of their resources to my pastoral setting. My project involves building a team of lay evangelizers and providing them tools to enhance their desire to engage inactive Catholics and the unaffiliated. Matthew Kelly's four signs of a dynamic Catholic—prayer, study, generosity, and evangelization—serve as a template for all the participants to develop habits that will enhance evangelization. Sherry Weddell's *Called and Gifted* discernment process is a strategy for team building. It reveals to the participants their gifts and offers them a renewed source of confidence to engage or evangelize the inactive Catholics and the unaffiliated.

I had two weeks of publicity to create awareness among parishioners and invited parishioners interested in being formed as evangelizers to sign up for the program. The buildup included a bulletin announcement inviting parishioners to participate in a formation program to teach them the skills and habits to be evangelizers. My invitation highlighted the activities participants will learn about to include: prayer, *lectio divina*, bible study, practices of generosity, discernment of their gifts or charisms, work as a team, and practices of evangelization based on their gifts.

The initial number of participants who signed up for the program was 25, not including the researcher. Next, I distributed to my volunteers an initial questionnaire to gauge their involvement within the parish. The questionnaire also enabled me to discern their familiarity with prayer, *lectio divina*, study, generosity, teamwork, awareness of their gifts, evangelization, and their use and knowledge of social media. I notified participants that this project is for a doctoral degree in preaching, and their feedback was significant in enhancing this project. After an initial orientation on the program's expectations, each participant signed an informed consent form to enable results to be published. After I reviewed the results from the questionnaires, I designed the weekly class sessions for the first six weeks to address the needs of the participants in the program. In the last three weeks, we launched our program for evangelization. Here is a brief overview of the process, which I will describe in more detail later on in this chapter.

In the first three weeks of the project, I offered instructions on prayer, *lectio divina*, study, generosity, evangelization, and the Creed. The design of these classes is to provide a refresher course on some of the subjects and introduce participants to the skills and knowledge that can enhance evangelization. From week four to week six, I introduced participants to the *Called and Gifted* discernment process to determine their gifts and how they can help them work as members of a team for evangelization. At the end of the three weeks of discernment and team building, members had the opportunity to develop an intervention to engage inactive members and the unaffiliated. In the final three weeks of the program, the participants engaged inactive members and the unaffiliated through our Facebook page designed for this outreach. Participants also exercised the skills developed throughout the formation program and engaged the target audience as teams within the parish context. They had the opportunity for three weeks to practice evangelization using social media.

5.3: Description of Participants

As pastor of Sacred Heart parish, I invited parishioners to participate in my thesis project designed to engage the Church's mission of evangelization targeting inactive Catholics and the unaffiliated. The selection of participants included a pulpit and bulletin announcement inviting all parishioners, whether cradle Catholics, reverts (Catholics who left and returned to the practice of the faith) or converts to Catholicism. The goal is to give all Catholics unaware of their responsibility to evangelize an opportunity to develop the knowledge and skills needed to evangelize.

In my selection of participants, I reminded parishioners that they would have an opportunity to learn new skills that will facilitate their desire to evangelize the parish's inactive members and the unaffiliated. Several parishioners who had seen the diminishing participation and the number of parishioners who had become inactive in the last ten years saw an opportunity to engage the inactive members within the community. I told the participants that the data from the project would be for publication. Also, I informed participants that the goal of my thesis was to support the laity's call and responsibility to evangelize.

In the Fall of 2018, while in the D.Min. program, I recorded my homilies for my professors and the class; this led to an interest in this form of communication. I decided to explore how I could engage my rural parish communities through the use of social media. I decided to take an elective on the theology of Social Media in the Spring of 2019 to develop the skills and expand my knowledge of social media engagement. While in the class, my professor challenged me to seek new ways of outreach to my rural parish community. I began to delve into social media for my weekly homilies, specifically online radio, YouTube, and Facebook. I have used social media as a form of outreach by engaging in a weekly devotional and sharing my weekly homilies. Throughout the time I have engaged in social media, it has been directed primarily to our parishioners in a forum entitled "From the Ambo." Many parishioners shared how these social media messages enhanced their participation and faith because it creates an opportunity for them to listen to the message at a different time and context outside of the mass. So, when I decided to develop a project using social media as a means of outreach to the inactive Catholics and the unaffiliated, some parishioners viewed it as being less judgmental and a possible way to engage the inactive members.

After my first meeting where I gave an initial orientation to prospective participants on the entire process and goals of the program, eight parishioners who signed up decided not to continue in the research project. The remaining seventeen parishioners stayed committed to the research project and persevered until its end.

All the participants are Caucasians and registered members of the parish. Of the seventeen participants, eleven were in their seventies, four were in their sixties, and two were in their fifties. Fourteen participants were female, and three were male. Twelve participants were cradle Catholics, while five were converts to Catholicism. Fourteen participants have been parish members for over twenty years, and three have been parish members for five years or less. Fourteen participants are married, two are separated, and one is a member of a religious community. Of the seventeen participants in the project, there were three couples. One of the couples participated through the online option. One participant is a graduate of the Rite of Christian Initiation in the past year. Fourteen participants chose in-person participation, and three participants chose online participation. The demographic of participants described aligns with the present membership at Sacred Heart Parish.

5.4: The Weekly Process

Week 1: Prayer and Study

In the first week of the program, I created a closed group on Facebook for participants in the formation program. Facebook allows online participants to be part of weekly sessions. The instructive and faith-sharing sessions were broadcast to parishioners from Immaculate Conception Parish, New Madrid and from Sacred Heart, Caruthersville to ensure that some feedback is received from persons outside of the program's participant-base. I began the session with an opening prayer and roll call to ascertain the number of online or in-person members in attendance. We held all the sessions for the formation program in a large meeting room in our parish school to facilitate social distancing and sufficient space for group discussions whenever necessary. Before introducing the topic of the evening, I invited participants to volunteer for various positions to simplify the process. The volunteer positions were: group secretary, group timekeeper, group social media director, and group accountant. Each participant who volunteered could participate fully in the program since his or her tasks required minimal attention while participating in the class sessions.

The volunteer secretary's task was to organize each participant's questionnaire or weekly survey after each class session. The volunteer group timekeeper's duty was to notify me of the end of the thirty-minute duration for each topic and to remind participants of the end of the ten-minute breakout sessions. The volunteer social media director's task was to help facilitate the online participant's presence by connecting and observing and sharing any comments, questions, or contributions from these participants. The volunteer project accountant's task involved recording any donations or financial contributions throughout the program.

I distributed an outline of the entire program to participants present and emailed copies to our three online participants. I went through the informed consent before distributing it to all the participants. I assigned the group secretary to compile a file for each person participating in the program to track all the handouts, surveys, and questionnaires distributed throughout the program. I proceeded to notify all the participants that the program would enable them to develop skills to be evangelizers and recognize their gifts while collaborating as teams to engage the inactive Catholics and the unaffiliated. I also informed the participants that the goal would be to increase their awareness of their call and duty to evangelize, engage inactive Catholics to increase their participation, and evangelize the unaffiliated to increase membership within the parish. Each introductory class session was at least thirty minutes, with ten-minute breakout sessions for discussions and questions from in-person attendees and online participants in the program. The discussion on prayer helped participants make a distinction between personal prayer and public prayer. Weekly instructional materials were available through handouts and PowerPoint presentations.

For the first week's instructional material, I started with a brief introduction from Matthew Kelly's book on the *Four Signs of a Dynamic Catholic*. I mentioned the four signs of a dynamic Catholic recommended by Kelly—prayer, study, generosity, and evangelization. I described how Kelly suggests at least ten minutes a day for prayer in the ten-minute rule.[175] However, participants were encouraged to use this rule in developing a habit regarding study, generosity, and evangelization at least once each day.[176]

I proceeded to share that for prayer to be effective there is the need to develop in attentive listening and quiet reflec-

---

[175] Matthew Kelly, *The Four Signs of a Dynamic Catholic* (Hebron, KY: Beacon Publishing, 2012), 57.

[176] Ibid., 60.

tion. I pointed out the distinction between private or personal prayer and public prayer. The difference was necessary to remind the participants that while our weekly gathering at the liturgy every weekend is an example of public prayer and an essential step to grow in prayer, our encounter with God is crucial for our personal growth in prayer. Public and private prayers complement each other in many ways and enhance our growth in faith, hope, and love as disciples. Before I ended this session on prayer, I invited participants to develop a habit of journaling their insights to facilitate personal accountability throughout their formation process. At the end of my introductory session on prayer, we went on a breakout session where participants could discuss insights and ask questions regarding prayer. In the breakout session, which was about ten minutes, I invited the online participants to post their comments to ask questions if they needed clarification on the material discussed in the session. The participants did not have any questions, but some observed their personal prayer experience has evolved over the years.

At the end of the breakout session, I introduced the next topic on study. On this topic, my primary focus was on scripture study and how it can affect our prayer life. I utilized Jeff Cavins *The Bible Timeline* as a primary resource.[177] While using this resource, I gave the participants an overview of various themes in the Old and New Testaments. After the outline of the Scriptures, I reminded participants that scripture study is essential for evangelization. During the breakout session, participants inquired about the difference between the Catholic and Protestant versions of the bible. I explained to participants that there were seven more books in the Old Testament than the Protestant Version for the Catholic version. I enumerated which additional books were present in the Catholic version of the bible. I took notes of questions and observed how participants engaged each other and me during the breakout sessions. Before the closing prayer, I invited participants to practice the ten-minute rule with prayer and study and prepare to offer feedback in the next session. I recommended that participants read a scripture passage for the ten-minute rule and then spend some quiet time reflecting on its application to their lives. I concluded the session with a closing prayer. The in-person attendance for the first week comprised 15 participants. There were five online participants and five parishioners who signed up and later decided to withdraw from the program.

---

[177] Jeff Cavins, Tim Gray and Sarah Christmyer, *The Bible Timeline: The Story of Salvation* (Milwaukee, WI: Ascension Press, 2011), 1.

Week 2: Catechism on Prayer and Scripture

In week 2 of the program, I started the session with an opening prayer. I offered a summary of the previous sessions and then invited participants to ask any outstanding questions regarding the sessions on prayer and study from week 1. I then proceeded to review *lectio divina* with the participants. I discussed the four-step process that has evolved from the monastic tradition. From the monastic practice, the four-step process is as follows:

- *Lectio* involves spiritual reading as foundational for an encounter with God.
- *Meditatio* involves reflecting, listening, and thinking of the words from the spiritual text.
- *Oratio* involves the vocal offering of prayer to God.
- *Contemplatio* involves expressions of love to God in prayer.[178]

The instruction on *lectio divina* was purely theoretical since ten of the participants had participated in a five-week class on the subject offered by a previous pastor of the parish.

---

[178] United States Catholic Conference of Bishops, *Catechism of the Catholic Church,* 2nd ed. (Washington, D.C.: United States Catholic Conference, 2000), 1177.

After a brief review of *lectio divina*, I turned my attention to *The Catechism of the Catholic Church*'s teaching on prayer.[179] I pointed out section and chapter one of the Catechism on prayer in the Christian life. I spoke on the sources of prayer and how to develop a method of prayer. I reminded the participants that, similar to *lectio divina*, spiritual reading is essential for growth in prayer. I also indicated that there are guides for prayer, and multiple ways of expressing prayer through devotion and pious acts. I noted the famous maxim that "anyone who sings prays twice." I offered some examples of spiritual reading materials that may assist participants in their prayer lives. The suggestions of spiritual reading texts comprised some books recommended by Kelly and some books that have contributed to my growth in prayer. Some of the spiritual reading materials suggested by Kelly that I shared with the participants include *The Story of a Soul: The Autobiography of St. Therese* and *An Introduction to the Devout Life by St. Francis De Sales,* among others. I recommended that participants read the following: *The Lives of the Saints* by Alban Butler, *Mere Christianity* by C.S. Lewis, or *Amazing Grace: A Vocabulary of Faith* by Kathleen Norris. After giving this summary on prayer from the Catechism, I paused for the first breakout session.

---

[179] *Catechism of the Catholic Church,* 2558-2758.

After the breakout session, I reviewed *The Catechism of the Catholic Church*'s teaching on Sacred Scripture.[180] In this follow-up session on the Scriptures, participants received a short review of the 46 books of the Old Testament and the 27 books of the New Testament. In my review of each book of both the Old and New Testament, I pointed out familiar stories in some of the books listed in the Old and New Testament. I reminded participants that while these bible stories are presented to us yearly in our liturgical celebrations, there is a need to have some personal time to reflect on these stories outside of the liturgy and engage their impact on our lives as Christians. At the end of my presentation, we had a breakout session. Participants had the opportunity to comment on the class and ask questions on biblical themes that needed clarification, like the Old Testament wars and Job's suffering. At the end of the breakout session, I invited participants to fill out two surveys to evaluate the sessions on prayer and study. I concluded the class with a closing prayer and adjourned for the week. The attendance for the second week comprised thirteen in-person participants and five online participants. Notable for this week is that the participants attending both online and in-person diminished, and instead of twenty like the previous week, we now had a total of seventeen participants.

---

[180] *Catechism*, 101-141.

Week 3: Generosity and Evangelization

I started the class session with an opening prayer. The first topic I instructed the participants on this week was *Generosity*. In this session, I reviewed that generosity is not limited to monetary possessions or personal wealth, according to Kelly's observation.[181] I also pointed out that when we talk about generosity, the typical focus is usually on our time, treasure, and talents.[182]

I shared that generosity is an essential part of our faith and is associated with stewardship. I pointed out to members that from its Latin root, it connotes a spirit of benevolence. Kelly notes that "generosity is at the heart of the Christian life, just as it is at the heart of the Gospel. For it is often through our generosity that we can bring the love of God to life for others in very tangible ways."[183] Therefore, understanding this sign of a dynamic Catholic requires an understanding of the concept of generosity. In the breakout session, participants considered some examples of generosity, including volunteering for various ministries and events within the Church. Each person's commitment to either an event or a ministry is an opportunity to practice

---

[181]Matthew Kelly, *The Four Signs of a Dynamic Catholic* (Hebron, KY: Dynamic Catholic Institute, 2012), 128.

[182] Ibid., 113.

[183] Ibid., 111.

generosity. As Kelly notes, "don't let what you can't do interfere with what you can do."[184]

After the breakout session, I reconvened the class for the second topic on evangelization. I introduced this topic by inviting the participants to watch a brief video by Bishop Robert Barron on *The Seven Great Qualities of a New Evangelist*. The seven qualities are

1. Love of Jesus Christ: Sharing a relationship of friendship with Jesus Christ.
2. New Ardor: Filled with ardor or fire for the Gospel.
3. Know the Story of Israel: Having knowledge and familiarity with the Old and New Testament story.
4. Know the Culture: A familiarity with the various generational changes in the society.
5. Love of the Great Tradition: Having a love of the Church's ancient tradition and values.
6. Missionary Heart: A missionary zeal and a commitment to daily participation in the Eucharist and a passion and hunger for saving souls.
7. Knows and Loves the New Media: Desire to embrace and utilize social media as a platform for engagement.[185]

---

[184] Ibid., 139.

[185] Robert Barron, *The Seven Great Qualities of a New Evangelist,* Web post accessed on 24 March 2021 at

After the nine-minute video, I used these points from Bishop Barron's insight to introduce the participants to what it means to be an evangelizing community. I took themes from Pope Francis' *Evangelii gaudium* and Pope Pius VI's *Evangelii nuntiandi* to highlight how each lay participant can be involved in evangelization. The themes I reviewed included how to be missionary disciples, how to be bold and creative, and how social media affects our lives as Christians. I concluded this session by reminding the participants that formation to be evangelizers is a process that requires their ongoing commitment and dedication. At the end of my thirty-minute class, I invited participants to discuss themes from Bishop Barron's strategy for new evangelists. Participants appreciated the breakdown of areas that a new evangelist needs to be aware of as he or she embarks on training and formation to be evangelizers. Many of the participants pointed out in our breakout session that having a missionary heart was important, especially when the outcome of our evangelizing effort does not yield the kind of positive response we expect.

At the end of the breakout session, I distributed two surveys to participants for the class on generosity and evangelization.

---

https://www.youtube.com/watch?v=w8EqX-bXF_I

Week 4: Introduction to Charisms

In the fourth week of the program, I introduced participants to the *Called and Gifted* discernment process to identify their gifts. This process aims to ensure that participants can recognize their gifts and develop as a team through this recognition with fellow participants who have similar gifts. The discernment process will facilitate team building and create opportunities for participants to grow in confidence as they collaborate with other participants to share their gifts. For members who find it challenging to share their faith with someone else, working in a team can help ease them into an increased engagement with someone else.

After the opening prayer, I asked participants to contribute $39 each for the program on *Called and Gifted*. Each participant contributed the amount required and received a booklet from the Catherine of Siena Institute: *Called and Gifted Discernment Process: Complete Resource* as part of the program. Besides the resource guide, participants received an email with a web link for each of the five sessions. Each session contained a series of short video segments that elaborated a gift. I invited the participants to recognize that a charism is

- A gift of the Holy Spirit given to every baptized Christian.

- A way of enabling Christians to accomplish things for God above and beyond natural human abilities.
- A special way that God's provision and love reach our neighbor and the world through us.[186]

The gifts to be discerned include twenty-three charisms. The first session of the booklet has five video segments. Some videos lasted about five minutes, and some lasted as much as twelve minutes. We watched all the video segments together, and at the end of session one, I invited participants to take the online inventory at home and then email their results to me. The online inventory comprises 120 questions that assist participants in discerning their core charisms.[187] The results from the online inventories were all determined by the experts at the Catherine of Siena Institute.

The topics in the first session, including the five video segments, are the following:

1. Introduction to the Called and Gifted Discernment Process

---

[186] Catherine of Siena Institute, *Called and Gifted Discernment Process: Complete Resource* (Colorado Springs, CO: Catherine of Siena Institute, 2019), 64.

[187] Sherry Weddell, *The Catholic Spiritual Gifts Inventory* (Colorado Springs, CO: Siena Institute Press, 1997), 60.

2. Introduction to Charisms and You!
3. Holy Spirit
4. Introduction to Charisms, Role and Gift Mix
5. How to Take the Inventory[188]

In the first video segment, the introduction outlines the goal of the discernment process and the learning expectation that "(the) parish is a source of evangelization, healing, and transformation for the world, and that your charisms and those of your fellow parishioners are a clue to the mission of your community."[189] The video also outlines the twenty-three core charisms highlighted throughout the process, which include:

1. Pastoral Charisms

- Encouragement
- Helps
- Hospitality
- Mercy
- Pastoring

[188] Ibid., 2-8.

[189] Ibid., 3.

2. Organizational Charisms

    - Administration
    - Leadership
    - Service
    - Giving

3. Charisms of Healing

    - Healing
    - Intercessory

4. Creative Charisms

    - Craftsmanship
    - Music
    - Writing

5. Communication Charisms

    - Evangelism
    - Prophecy
    - Teaching

6. Lifestyle Charisms

    - Celibacy

- Extraordinary Faith
- Missionary

7. Voluntary Poverty

- Charisms of Understanding
- Knowledge
- Wisdom[190]

The second video segment introduced what it means to be a disciple and the importance of a relationship with Jesus, cooperating with the Holy Spirit, and when and why we receive the gifts.[191] The third video segment focused on how the Holy Spirit comes into our lives through the graces we receive at baptism, confirmation, and Holy Communion.[192] The segment ends with a reminder of the obstacles to the Holy Spirit, including a lack of faith and a lack of repentance.[193] The fourth video segment on the role of charisms and the combination of gifts in our lives as disciples focused on actions resulting from our Christian lives and how we all receive more than one gift and their effects on our lives.[194] The fifth video segment was a review of the in-

---

[190] Ibid., 12.

[191] Ibid., 4.

[192] Ibid., 6.

[193] Ibid.

[194] Ibid., 8.

ventory and instructions on how to take the inventory.[195] I invited the participants to email their scores to me after receiving their results. At the end of the session, I asked participants to watch the video segments for session two at Home and note any questions for the next class. I concluded the session with a closing prayer.

Week 5: Discernment of Charisms

After an opening prayer, I informed participants that I received most of the scores via email. Participants who had not emailed their scores to me submitted their online inventory scores in person. I went through all online inventory scores received from the participants and shared the top six scores of every participant. I informed the participants that the top six scores highlight their most substantial gifts. The review of the scores of each participant present and the online participants lasted about two minutes for each participant. Each participant had about five minutes to react to the online scores. Most participants affirmed that the charisms reflected in the highest score from the online inventory were their strongest. Three participants indicated that they wanted to discern their second-highest charism to confirm its presence in their lives.

[195] Ibid., 9.

I invited each participant to engage in a personal discernment in the coming week through interaction with other participants, a family member, or a friend to confirm if the gifts highlighted by the inventory truly reflect their giftedness. The discernment process involving at least one family member, associate, or friend will enable the participant to receive initial feedback on their chosen charism. Following input from family members or friends, the next step is discerning with a group to receive confirmation or different feedback. From the booklet for the discernment process, I directed the participants to engage in the following steps for their discernment in the next week:

1. Our personal experience:

    - Feelings of satisfaction/joy; it is energizing
    - A sense that we "fit" or are "in our place."
    - An expression of our relationship with God
    - Like prayer or contemplation

2. Our effectiveness:
    - We are a channel of God's love and provision for others.
    - We are remarkably effective without undue struggle.

3. Feedback from others:
   - Direct feedback
   - Indirect feedback[196]

I invited participants to journal their experiences as this will enable them to confirm the gifts highlighted in the inventory results.

After reviewing the online inventory and the next steps for their discernment, I invited the participants to share their observations from session two, which was assigned the previous week. Participants raised questions about the insight they received when one of the video segments highlighted the difference between a natural talent that one inherits and a charism. I also mentioned that session two in our booklet mainly focuses on the pastoral charisms—encouragement, helps, hospitality, mercy, and pastoring—and these charisms facilitate a "personal *nurturing* of individuals and community."[197] I invited the participants to watch a few video segments from sessions 3, 4, and 5 of the *Called and Gifted Discernment Process*, and we watched these video segments together in the class. In the video segments, I reviewed with the participants the following topics:

---

[196] Ibid., 64.

[197] Ibid., 13.

- Communication Charisms: these are charisms that enable a person to transform lives by communicating truth.[198]
- Organizational Charisms: these are charisms that enable a person to meet "the vision or structural needs of an organization."[199]
- Lifestyle Charisms: these are charisms that "empower a distinctive lifestyle and freedom for unusual ministry."[200]
- Personality and Discernment: these factors may affect our exercise of the charisms and how we embrace the charisms in our lives. For instance, extroverts are prone to take actions faster while over-estimating their giftedness, and introverts act more slowly while under-estimating their giftedness. [201]
- Healing charisms: these are charisms that enable a person to be "a channel of God's healing and restoration."[202]
- Charisms of Understanding: these are charisms that enable a person to understand "the ways of God and the universe."[203]

---

[198] Ibid., 18.

[199] Ibid., 22

[200] Ibid., 29.

[201] Ibid., 3

[202] Ibid., 33.

- Creative Charisms: these are charisms that "inspire creativity with material things – paint, clay, wood, food, plants, cloth, digital media, etc."[204]

After reviewing these segments, in the breakout session, I reminded the participants to journal their experiences from the discernment process as a way of helping them recognize how God desires to use them in reaching out to a family member, an inactive Catholic, or an unaffiliated person.

At the end of our discussion on the video segments, I invited participants to complete a survey I had developed to evaluate their experience on the online inventory and the class session. I ended this session with a closing prayer.

<u>Week 6: Team building</u>

After participants had gathered in our class this week, I opened the session with an opening prayer. I then invited in-person participants or online participants to share their experiences thus far on the discernment process. The online participants had the option of calling in if they wanted to share with the class their experiences. A few participants shared how empowering it was to know their gifts

---

[203] Ibid., 37.

[204] Ibid., 38.

and how it boosted their confidence as they recognized that other participants had similar gifts. While they knew they had certain gifts, participants shared that receiving feedback from family members or friends affirmed their gifts. Through the discernment, participants recognized that charisms are a favor from God that is "always for the sake of others" and "make us instruments of his providence."[205] After a few participants had shared their experiences, I informed them that I would allocate the teams based on the similarity of gifts. I shared that working with another participant with a similar gift will enable members to grow in confidence as they work together. Also, there will be easier collaboration within a group of similar gifts as members work to develop as a team. Teamwork will facilitate the success of our intervention for evangelization and engagement of the inactive members and the unaffiliated.

All the participants agreed to this, and I proceeded to call out each participant and their strongest charism highlighted in the online inventory. I also assigned three participants who chose to discern their second-highest charisms into their respective teams, reflecting their selected charism. From the online inventory and the discernment process, the following groups emerged as our teams: three participants had charisms of encouragement, four participants had organizational charisms, five participants had hospi-

---

[205] Ibid.

tality charisms, three participants had intercessory charisms and two participants had creative charisms. After the grouping, I invited the class to take a five-minute break so that participants could recognize who was in their assigned team.

Since the charism of encouragement classification is part of the pastoral charisms, I reminded the three team members that their primary task is to facilitate the personal nurturing of individuals and the community.[206] Participants with the charism of encouragement "see individuals through the eyes of God and encourage them to take their place in the Kingdom."[207] I reminded the encouragement team about the goal of our evangelization outreach to invite and encourage the inactive members and the unaffiliated to an encounter with Jesus Christ in the Church. After the class, the team members suggested a meeting to study and propose three scriptural passages that will enhance their engagement of the inactive Catholics and the unaffiliated. The entire class agreed to this proposal.

I assigned participants with high scores in administration, leadership, service, and giving for the organizational charism team. These participants had the task of helping organize the group through their specific charisms. From our brainstorm, members of this team decided that for our

---

[206] Ibid., 13.

[207] Ibid.

online engagement, everyone in the group should come up with names of members within the parish community who have not been to Church for at least three weeks or beyond. A Facebook invitation will be sent to members on the list to join our live stream evangelization sessions. The participants all agreed to this proposal.

I assigned the participants with high scores in hospitality to the hospitality team. I informed the team members that their gift "is one of the primary means by which God heals and strengthens individuals through the experience of Christian community."[208] Through our brainstorming, the hospitality team proposed a regular mail invitation to the inactive members and the unaffiliated for an in-person visit, follow-up, and welcome to the parish. The participants all agreed to this proposal.

For the intercessory team, I assigned participants who had high scores in intercessory prayer. I reminded the team members that through their gift, "they see remarkable, even miraculous, answer to their prayers for others regularly. They are often directed specifically in their prayers by the Holy Spirit, and they find praying for others energizing."[209] Through our brainstorming, these participants indicated that they would take it upon themselves to meet once a week to pray for the return of all inactive members and for

---

[208] Ibid., 15.

[209] Ibid., 34.

openness to listen to the invitation to return to active participation. All the participants accepted their proposal.

Finally, I assigned two members to the creative charism team. I reminded the team members that "through craftsmanship, music, and writing (the charism) inspires the baptized person to be creative with material things like paint, clay, wood, food, and digital media."[210] The team members proposed to write short prayers for use at our gatherings to engage the inactive or unaffiliated persons. All participants accepted the proposal. One of the creative team members also volunteered to assist with the social media duties of the class throughout the live stream. After this brainstorm with each team, I invited the participants to take a ten-minute break and reconvene after the break to discuss the outline of our evangelization intervention. During the break, participants began to brainstorm by suggesting names of inactive members. Some participants suggested family members who had been absent from the parish community for over a year. Since Sacred Heart parish is a relatively small Catholic Community, many participants knew the inactive members. As members made suggestions of inactive members, a member of the organizational team noted the names. After about fifteen minutes, I invited the organizational team member to read out the names of inactive members on the list. As members heard the names,

---

[210] Ibid.

there were a few more suggestions. After another five minutes, the suggestion of names of the inactive members within the parish boundaries numbered about fifty-six persons.

I invited all the participants to brainstorm to develop an outline that we could use for our evangelization outreach. Since Facebook allows a wide range of following and allows viewers to post comments or indicate their reactions to the contents of the page, the five teams agreed on the following guidelines using the Facebook live stream as our primary means of outreach to the inactive members and the unaffiliated:

1. Welcoming online viewers by a member of the organizational team [2 minutes]
2. An invitation to prayer by a member of the creative team [2 minutes]
3. Reading of the scriptural passage for the day by a member of the encouragement team [2 minutes]
4. A brief scriptural reflection by each member of the encouragement team [3 minutes for each reflection] followed by about two minutes of quiet time [15 minutes].
5. A summary of the reflection by a member of the organizational team [2 minutes].
6. An invitation to the viewers to post their comments or reactions to the reflection [2 minutes]

7. A member of the hospitality team then offers an invitation to all viewers to join in our weekly celebration of the mass/ adoration [5 minutes]
8. A member of the intercessory team shares their support and prayers for all viewers [5 minutes]
9. Closing remarks by an organizational team member [2 minutes]
10. Invitation to prayer by a member of the creative team [2 minutes]

After we discussed the outline of the intervention, participants suggested that the entire outreach session should be within thirty or forty minutes. As a framework for the reflections for the viewers, I invited the encouragement team members to focus on how they can reassure the inactive members of the parish's concern and hope that they return to active participation. I also challenged the encouragement team that their reflections should invite inactive members back to the community and remind them that they are still a part of the parish community. The goal of the intervention is to offer the inactive members and the unaffiliated persons a welcoming community that reaches out with the message of Jesus Christ. The reflections should be short and directed to the intended audience—the inactive members and the unaffiliated. The time limit would be three to five minutes, and the viewers will have an opportunity to reflect on the words heard for two minutes. After the sharing, the hospitality and intercessory team members will reassure the viewers of our ongoing prayers for their return and notify views of activities within the parish. The organizational team members will summarize the messages to reinforce that their presence in our community is essential and offer the parish phone number and mailing address for any comments and follow-up. Through our discussion, I assigned the creative team members to write out a prayer we could adopt for our evangelization outreach sessions, and they agreed to develop one for the sessions. All participants agreed to this format. I reminded everyone present that after I review the scripture passages selected by the encouragement team, I will email them to all participants to read in preparation for our evangelization outreach. Before the concluding prayer, I notified the organizational team members to contact all inactive members listed for our evangelization outreach and invite them to follow our Facebook page and share the date and time for our outreach to them. After discussing an appropriate time, participants agreed that our outreach would be at 5:30pm every Wednesday. The designated time will allow persons in our target audience who have a work schedule the opportunity to join and follow our live stream. I asked members to inform all persons invited that their online presence and feedback through comments posted, call-in made to the parish office, or email to our parish will be appreciated. I also informed participants that the evangelization sessions would be open to all members of the parish community. I concluded the session with a prayer, and participants departed.

Testing and Evangelization

Week 7

Before the evangelization reflections and activities began, the encouragement team contacted me with the following passages for the last three weeks of the program for their reflections and outreach.

- Week 7 reflection passage: Matthew 5: 14-16
- Week 8 reflection passage: Hebrews 10: 24-25
- Week 9 reflection passage: 1 Peter 3: 15-16a

The evangelization outreach goal includes inviting, welcoming, and encouraging the inactive members and the unaffiliated to encounter Jesus Christ in the Church. From my discussion with the encouragement team, the first passage, *Matthew 5: 14-16,* was a recommendation from the resource book for the *Called and Gifted* discernment.[211] The team members indicated that from several passages they studied about evangelization and encouragement, they

[211] Ibid., 73.

agreed that *Hebrews 10: 24-25* and *1 Peter 3: 15-16a* were suitable for our context. After approving the passages, I emailed all the participants the selections and invited everyone to read the passages before our weekly meetings.

In week seven of the program, my role was primarily to facilitate outreach to our target audience—inactive members and the unaffiliated. A member of the creative team was available to report any online comments from our viewers. We also made our parish phone number available to our viewers who may need to call in to contact us. After setting up via Facebook, we opened with a prayer written by a creative team member. The prayer was displayed on the screen for all to pray along. After the opening prayer, I introduced the five teams to our audience and briefly described their duties. After my introduction, I informed our viewers of the passage for the evangelization outreach.

> You are the light of the world. A city set on a mountain cannot be hidden. Nor do they light a lamp and then put it under a bushel basket; it is set on a lampstand, where it gives light to all in the house. Just so, your light must shine before others, that they may see your good deeds and glorify your heavenly Father (*NAB*, Matt. 5:14-16).

After the passage from Matthew 5:14-16 was read, I invited each member of the encouragement team to share their reflection on the passage. At the end of each reflection, I asked viewers to reflect briefly on what was heard and post their comments. The recurring theme of the reflection was on "being a light for others," and as regards our inactive members, it is inviting them to an encounter with the light, Jesus Christ. Some viewers posted comments appreciating the presentation, and some posted that they liked the reflections. There were no extensive comments but acknowledgments of the efforts of the entire team.

After the reflection, I invited a member of the hospitality team to share about any upcoming events that will facilitate an in-person encounter with any interested listener from our target audience. The team member highlighted a future event, and I proceeded to ask a member of the intercessory team to reassure our viewers of their commitment to pray for them as they discern a possible return to active participation in the parish. At the end of the intercessory member's comments, I reminded our viewers of the following weeks. I shared our parish phone number and email address for any comments and follow-up questions or concerns. At the end of my closing remarks, I invited all—viewers and participants—to join in the concluding prayer displayed on the screen.

At the end of the live stream, I invited participants to review the initial evangelization outreach process to see beneficial areas and areas that need a different approach. Participants suggested that the repetition of our contact information was a good idea so that viewers could contact the parish. I invited the organizational team members to assign one of the members to facilitate the subsequent outreach sessions, and all accepted this.

Week 8

In week 8, participants gathered, and we set up for the live steam. I invited a member of the organizational team to facilitate the session. I stayed with the participants in the room to direct the teams according to our guidelines for the outreach. The passage for this week's outreach was Hebrews 10: 24-25.

The organizational team member invited who was facilitating the process called on a member of the encouragement team to read the passage.

We must consider how to rouse one another to love and good works. We should not stay away from our assembly, as is the custom of some, but encourage one another, and this all the more as you see the day drawing near (*NAB*, Heb. 10: 24-25).

After reading the passage, all three members each gave a brief reflection on the passage followed by a few minutes of silent reflection. The facilitator invited the viewers to post their comments and reactions to the reflections and reminded them of our phone number for call-in. The general theme for this week's reflection is that "we need to encourage each other and build the body of Christ together." Some viewers posted comments appreciating the presentation, and some posted that they liked the reflections. The facilitator then invited a hospitality team member to remind viewers of the upcoming in-person encounter for a follow-up to assist interested persons who desire to return to participation with the parish community. After the hospitality team member, the intercessory team reminded the viewers of the team's prayers for them. The facilitator invited a creative team member to lead with the concluding prayer.

Like the previous week, the participants stayed back and reviewed the day's session to see areas that needed to be sustained and improved for the subsequent week's presentations. Aside from the encouragement and creative teams, the other teams appointed the members responsible for the presentations in the coming week.

Week 9

In week 9, after we set up for the live stream, I invited a member of the organizational team to facilitate the session like the previous week while I joined the participants in the room to observe them throughout the process. The passage for this week's outreach was 1 Peter 3: 15.

The organizational team member called on a creative team member to lead in the opening prayer displayed on the screen. The facilitator, after the prayer, invited a member of the encouragement team to read the passage.

> But sanctify Christ as Lord in your hearts. Always be ready to give an explanation to anyone who asks you for a reason for your hope, but do it with gentleness and reverence (*NAB*, 1 Pet. 3: 15-16a).

After reading the passage, all three members participated in a few minutes of silent reflection after each sharing. The facilitator then invited the viewers to post their comments and reactions to the reflections and reminded them of our phone number for call-in. The theme for this week's reflection is that "sharing the hope we have in Jesus makes the difference in our relationship with others." Some viewers posted comments appreciating the presentation, and some posted a 'like' symbol. The facilitator then invited a hospitality team member to remind viewers of the in-person encounter. The hospitality team member reminded the viewers that will be receiving invitations for the gathering in the parish. After the hospitality team member's announcement, the intercessory team member reminded the viewers that there was a prayer chain to accompany them with prayers as they discern a possible return to the parish community. The facilitator invited a creative team member to lead with the concluding prayer at the end of these comments.

At the end of this session, I thanked all the participants for their dedication and distributed a survey. I reminded them that the survey would assist me in determining what aspects were beneficial and what parts needed to improve. The hospitality team and the organizational team members shared the date of the in-person follow-up of inactive members and the unaffiliated and reminded all participants to be present.

A week after the program's end, fifty-six inactive members who were invited to follow the online faith-sharing engagement were invited for an in-person follow-up. Four persons showed up and all participants in the evangelization program were present in the parish hall to welcome them and share a meal with them. At this gathering, inactive members were invited to join at any of the weekend masses and the hospitality team members assured them of their presence to accompany them through the process of returning to active participation.

5.5 Data Collection and Conclusion

The data collected for the program was through surveys distributed throughout the process. Participants also received an initial questionnaire to help assess their level of involvement within the parish and their knowledge on prayer, study, generosity, evangelization, the discernment of gifts, and teamwork. Besides the surveys, one participant took notes of all discussions, questions, comments, and observations proposed through the nine weeks of the program. Another participant was also appointed to track the weekly online participation and viewers via Facebook live stream. At the end of the program, I had interviews to access participants' perspectives and sent out surveys to viewers of the evangelization outreach. Throughout the process, I took notes for my observation of the reactions to the program. The surveys, interviews, and personal notes are the data I will review in the next chapter.

Therefore, to facilitate proper triangulation in this project, the perspectives that will guide the results and recommendations in the subsequent chapter will be derived from the sources involved throughout the intervention process. These perspectives include the participants who engaged in learning throughout the program; their written responses to the surveys, questionnaires, and interviews; and the written feedback from observers, the inactive members, or the unaffiliated who followed us throughout the Facebook live stream outreach, and my notes and observations.

Chapter 6

Findings and Next Steps

6.1: Introduction

In the previous chapter, I discussed the intervention strategy and weekly sessions of my thesis project. I also discussed the participants involved in my project and the mode of data collection. Included in these considerations was a brief discussion on the *Called and Gifted* discernment process that enabled participants to discern their charisms and how these charisms enhanced their preparation and participation in evangelization. I also noted how the discernment of gifts facilitated team assignment and team-building among participants.

In this chapter, I will be reviewing the results from the questionnaires, surveys for the weekly sessions, my notes, and interviews conducted to gauge participants' feedback and overall experience. All nine weeks of the program were live streamed and there was a final intervention of inviting the inactive and the unaffiliated to the parish. Besides the participants' feedback, I will review the feedback from the inactive Catholics and an observer from another parish within the bootheel. These will provide input as to the effectiveness of the project in strengthening members of the laity to engage inactive Catholics and the unaffiliated. Also,

I will be discussing the impact of social media on my ministerial intervention and its effects on participants throughout the program.

In this chapter, I will rely on Tim Sensing's recommendation on triangulation to facilitate proper interpretation. Sensing writes that "the key to interpretation is not in the tools, or the reporting of the information guided by a set of techniques, but rather in how one makes sense of the experience of everyday life."[212] With this consideration, I will discuss how the feedback from questionnaires, surveys, and interviews provides information on how the participants' grasp of evangelization's basic knowledge and skills evolved throughout the process. I will also review some of my observations from my field notes throughout the process. These observations will factor into my recommendations for the program's future and how the program could contribute to evangelization in southeast Missouri, in the diocese of Springfield-Cape Girardeau, and other rural dioceses in the United States.

---

[212] Tim Sensing, *Qualitative Research: A Multi-Methods Approach to Projects for Doctor of Ministry Theses* (Eugene, OR: Wipf and Stock Publishers, 2011), 72.

6.2: Data Analysis

At the start of the nine-week program, there was initial excitement and anxiety among parishioners on what to expect. The anticipation of not knowing the outcome was undoubtedly felt by me—the researcher—as I approached the project. Members of the parish responded with enthusiasm while signing up for the program on evangelization and the possibility of learning something new and developing a way to reach out to inactive members and the unaffiliated. In the subsequent weeks, this initial enthusiasm strengthened members who participated throughout the nine weekly sessions. The feedback and observations received help contribute to enhancing the development of the program and supporting the participants in evangelizing the inactive and the unaffiliated.

In my first chapter, I indicated that the problem that my thesis will address is that the parishioners are unaware of their call to evangelize. Also, I noted that some parishioners have reservations about the parish's future due to diminishing attendance and not knowing how to address the issue. These reservations about the parish and possible steps for the future led to discussions about how to evangelize and establish a process to engage inactive members and the unaffiliated to build up parish membership and participation.

To address the lack of evangelization and provide training and resources to increase participation and membership within the parish, I chose to form lay participants in the following areas; prayer, study, generosity, evangelization, and the discernment of their gifts. Based on my research, the works of Matthew Kelly and Sherry Weddell were essential for my project. Also, regarding applying these skills and knowledge through social media, I chose Facebook live stream as a primary means of communicating to our target audience.

The program was a learning opportunity to instruct the participants and assist in developing skills that will encourage a commitment to evangelization within our rural parish community. I distributed questionnaires before the start of the program, and at the end of the nine-week program, I interviewed the participants to gauge their experiences. Seventeen members participated in the entire nine-week program and responded to the questionnaires before the program. At the end of the program, I interviewed the participants on questions concerning the core themes and skills in each weekly session.

The week before the program's start, I invited members to fill out and respond to questions concerning parish membership and involvement. In the questionnaire, I also asked participants to respond to various practices contributing to their growth within the Church. Finally, I invited participants to indicate their familiarity and previous training or knowledge of prayer, study, generosity, evangelization, team building, discernment of charisms, and social media training. Of the six questions in the questionnaire, four questions required a direct response. In the chart below, the participants' responses show that most participants in the nine-week program have been in the parish for over twenty years.

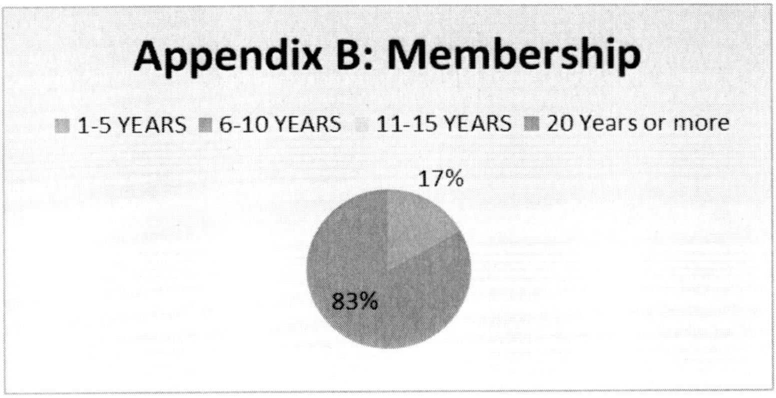

Figure 6.1: Responses from participants on their years of membership within the parish.

The participants responded to their level of involvement in the parish, and the chart below shows that a majority of the participants are very involved within the parish. Participants' involvement within the parish here indicates involvement in various ministries such as lectors, sacristans, Eucharistic ministers, choir members, greeters/ushers. The involvement of these parishioners also includes their membership in parish sodalities like the women's group, the rosary group, or other prayer and social groups within the parish.

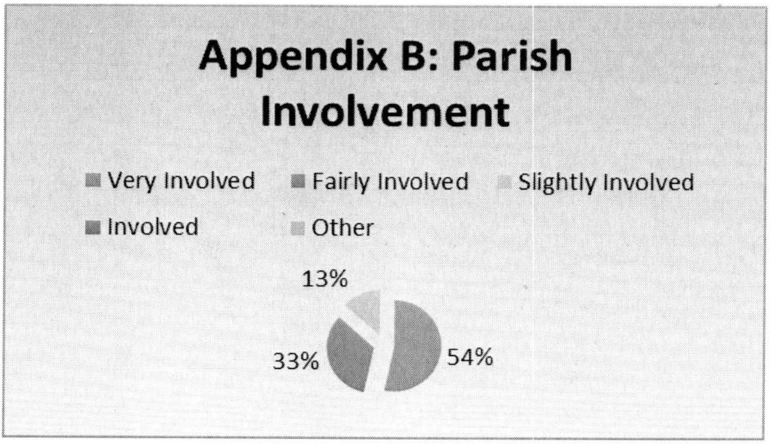

Figure 6.2: Participants' responses to their involvement within Sacred Heart Parish.

In the questionnaire, I invited participants to respond to what aspects of life in the parish have contributed to their growth. To this question, the responses varied, as the chart below reveals. The parish community, personal prayer, and the mass and reconciliation were the top three factors from the participants' responses.

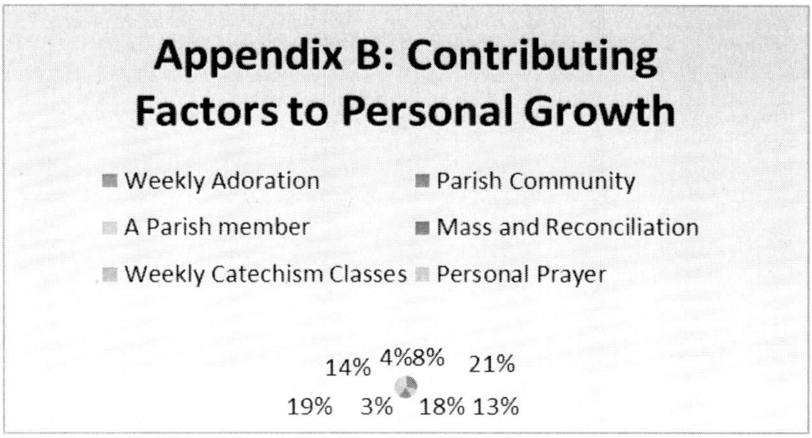

Figure 6.3: Participants' responses to factors contributing to their ongoing growth within the parish as Catholics.

I asked participants to indicate what areas they have received training or formation as it relates to the class; 9 indicated a prior training in *lectio divina*, 2 stated that they had some formation in the discernment of their charisms, 3 indicated that they had some training in team building, 1 participant had some experience of evangelization, and 2 had some prior training in social media use and engagement. The chart below outlines the result from participants' responses in percentage points.

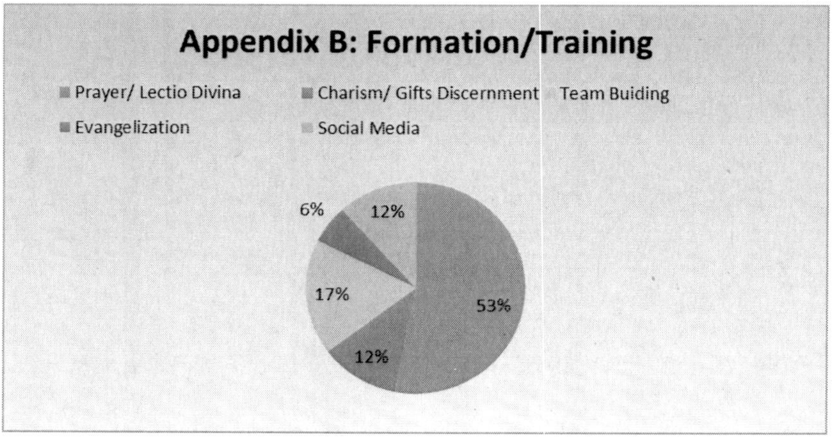

Figure 6.4: Participants' responses to their familiarity with significant themes and skills before the class.

The chart highlights that going into the nine-week program, more than half of the participants had some knowledge of *lectio divina* as a spiritual practice from previous training. The response shows that prayer and *lectio divina* were the most vital link as the participants begin their formation and evangelization in the area of least experience and training.

As the program progressed, the weekly surveys assisted me in evaluating the participants' learning and progress concerning the themes and skills throughout the nine-week program. Each weekly survey often came a week after a class on the specific topic or skill to enable participants to reflect, practice, and evaluate their development. I made some notes as the weeks progressed to observe how participants evolved throughout the program. I will summarize the participants' learning responses to the surveys in the breakdown below.

After weeks one and two, participants responded to questions on prayer and study. The surveys comprised a list of statements to rate their learning on these spiritual practices. The questions on the surveys were to rate their level of agreement with "strongly agree" as the highest level and "strongly disagree" as the lowest level. The chart below displays the participants' responses to their learning on the class on prayer. The question about the presentation being easy to follow and having a clear idea; 12 strongly agreed, and 5 agreed. For mental imagery and examples appealing to a novice in prayer, 5 strongly agreed, 10 agreed and 2 slightly agreed. For new information that challenged their approach to prayer, 11 strongly agreed, 3 agreed, and 3 slightly agreed.

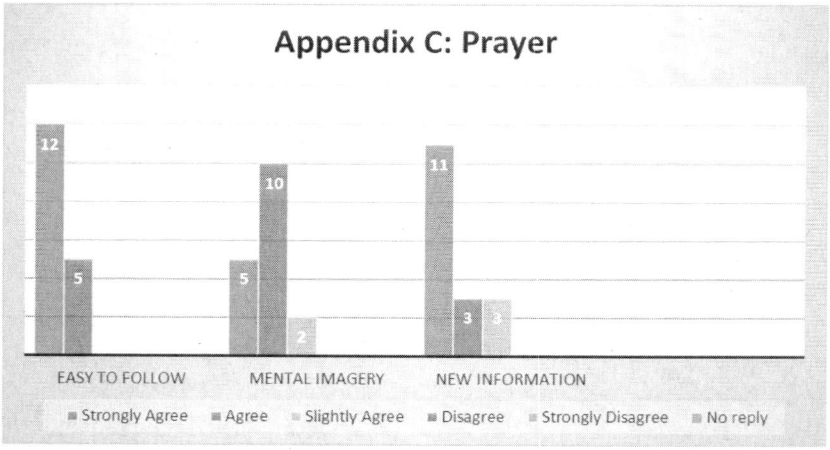

Figure 6.5: Participants' responses to questions on ease in following class, mental imagery, and new information.

On needing more sessions on prayer, 5 strongly agreed, 11 agreed, and 1 strongly disagreed. The session helps distinguish between private and public prayer; 9 strongly agreed, 7 agreed, and 1 slightly agreed. In response to whether the mass is helpful preparation for evangelization, 5 strongly agreed, 11 agreed, and 1 slightly agreed.

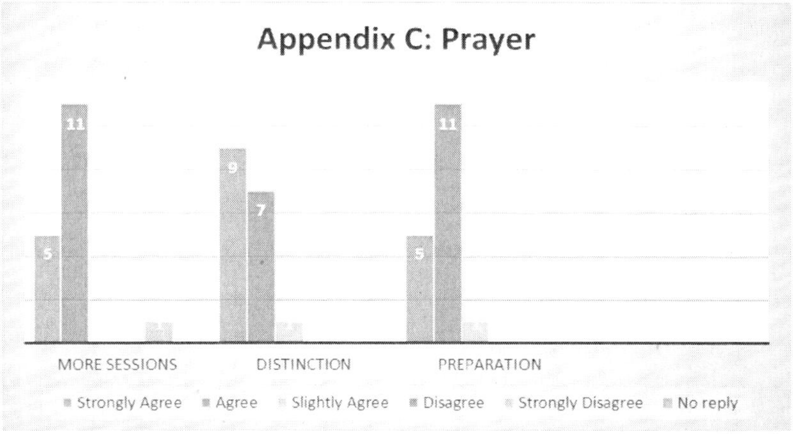

Figure 6.6: Participants' responses to questions on more sessions, distinction, and preparation.

On the 10-minute rule enhancing their prayer life, 5 strongly agreed, 8 agreed, and 4 slightly agreed. Regarding praying the rosary being important in their preparation for evangelization, 4 strongly agreed, 10 agreed, and 3 slightly agreed. The next question was on recommending the class on prayer to someone desiring knowledge and growth in prayer; 11 strongly agreed, 5 agreed, and 1 slightly agreed.

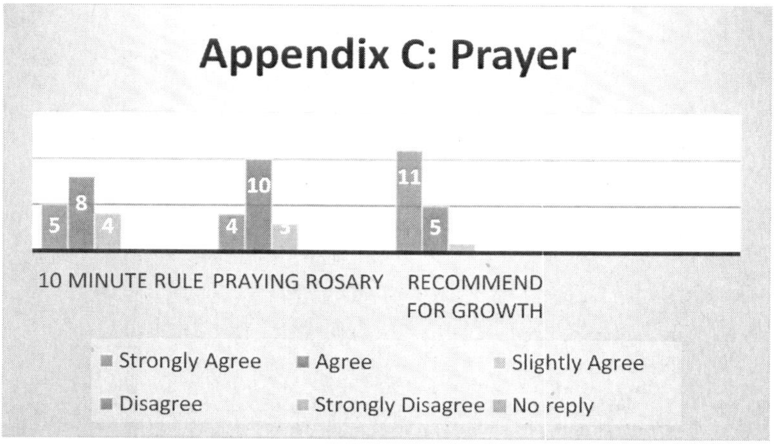

Figure 6.7: Participants' responses to questions on the 10-minute rule, praying the rosary, and recommending the class to someone desiring growth in prayer.

On using social media in this session to enhance their learning, 5 strongly agreed, 7 agreed, 4 slightly agreed, 1 gave no reply. In response to recommending the session to someone desiring to participate in evangelization, 4 strongly agreed, 12 agreed, and 1 gave no reply. The overall response on the class on prayer indicates that the participants were engaged and found the class on prayer beneficial for their training for evangelization; however, while most participants indicated a need for more sessions, one participant strongly disagreed with the need for more sessions.

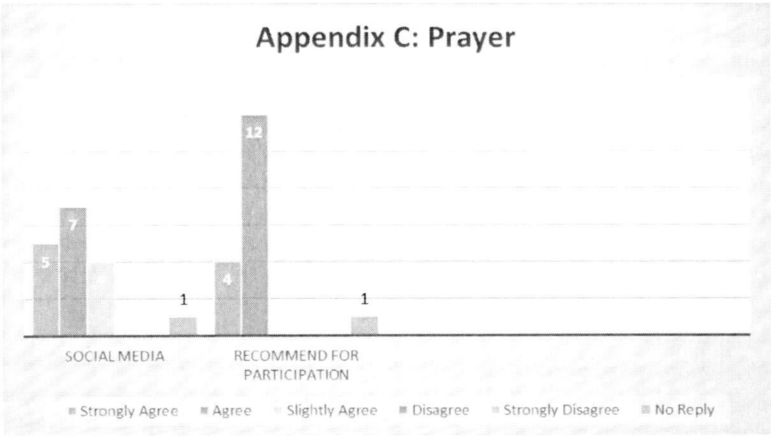

Figure 6.8: Participants' responses to questions on social media use and recommending the class for preparation for evangelization.

At the end of the second week, participants provided their feedback via a survey on the class on study. In response to the question, the session had a clear idea that was easy to follow; 4 strongly agreed, and 13 agreed. To the question that "the session had examples and mental imagery that enhanced my formation," 3 strongly agreed, 13 agreed, and 1 gave no reply. In response to the question that "the session had new information that challenged my approach to study," 5 strongly agreed, 9 agreed, and 3 slightly agreed.

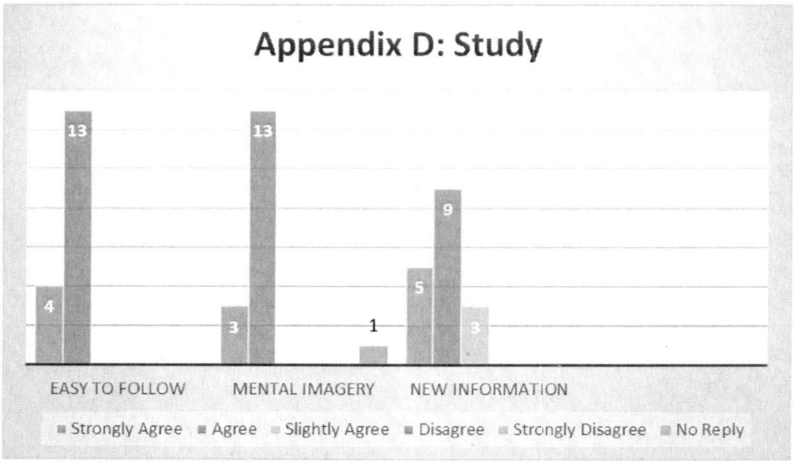

Figure 6.9: Participants' responses to questions on ease in following the class, mental imagery, and new information.

On the need for more study sessions, 3 strongly agreed, 10 agreed, 3 slightly agreed, and 1 strongly disagreed. In response to the question that "the session helped me distinguish between spiritual reading and regular reading," 7 strongly agreed, and 10 agreed. Regarding the question that "participating in regular study and spiritual reading is a helpful preparation for evangelization," 5 strongly agreed, and 12 agreed.

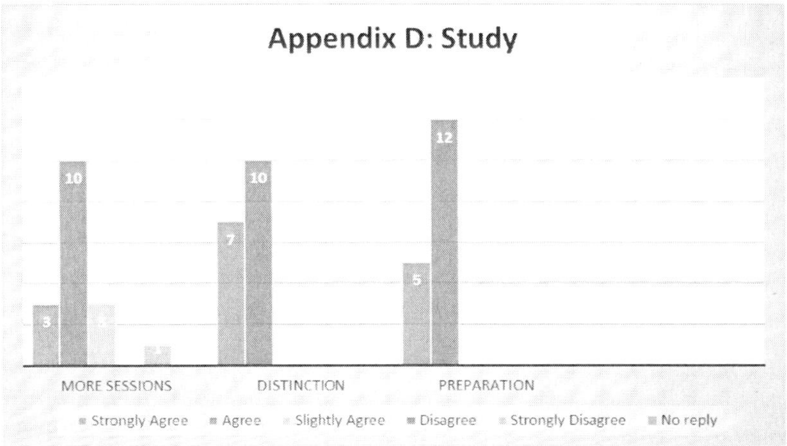

Figure 6.10: Participants' responses to questions on more sessions, distinction, and preparation.

On the 10-minute rule for study each day, 5 strongly agreed, and 12 agreed. Regarding scripture study as helping prepare participants for evangelization, 8 strongly agreed, 8 agreed, and 1 slightly agreed. The next question recommended the class on study to someone desiring knowledge in Church teaching; 10 strongly agreed, and 7 agreed.

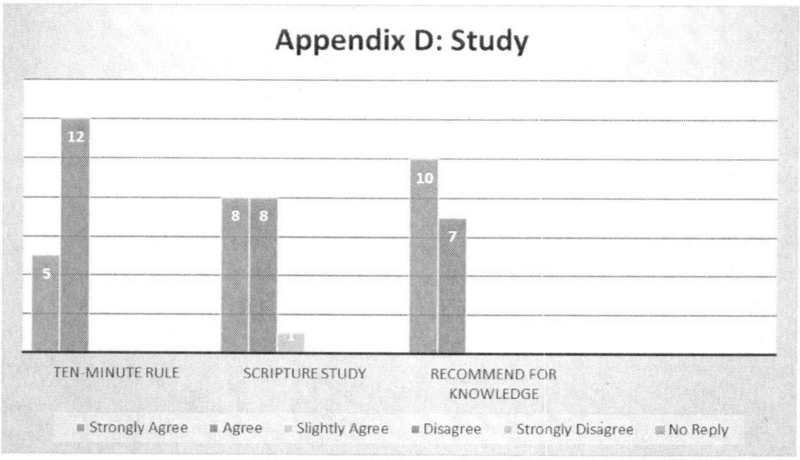

Figure 6.11: Participants' responses to questions on the 10-minute rule, scripture study, and recommending the class to someone desiring knowledge of Church teaching.

On using social media in this session to enhance their learning, 2 strongly agreed, 11 agreed, 3 slightly agreed, and 1 gave no reply. Regarding recommending the class for preparation for evangelization, 7 strongly agreed, 7 agreed, 1 slightly agreed, and 2 gave no reply. The overall response from the class on study indicates that the participants were engaged and found the class on study beneficial for their training for evangelization; however, while most participants indicated a need for more sessions, one participant strongly disagreed with the need for more sessions.

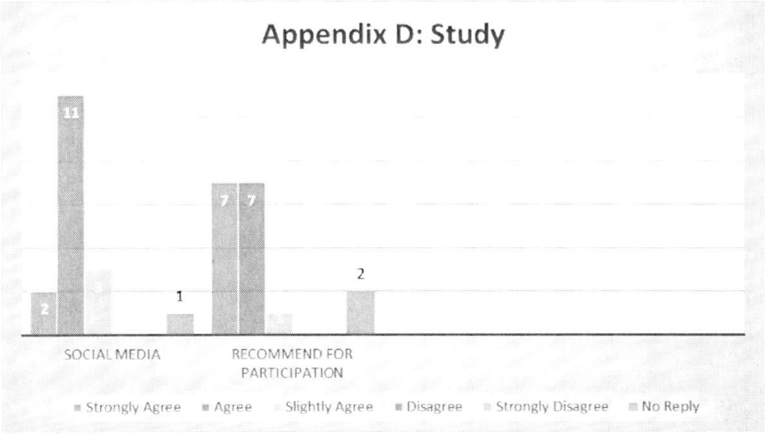

Figure 6.12: Participants' responses to questions on social media use and recommending the class for preparation for evangelization.

At the end of the third week, I asked participants to give feedback on their learning on generosity and evangelization. The charts below indicate the response of participants for the class on generosity. In response to the question, the session had a clear idea that was easy to follow; 11 strongly agreed that it was easy to follow and had a clear idea, and 6 agreed on the same question. The question that "the session had examples and mental imagery that enhanced my formation;" 4 strongly agreed, and 13 agreed. In response to the question that "the session had new information that challenged my approach to generosity," 5 strongly agreed, 8 agreed, and 4 slightly agreed.

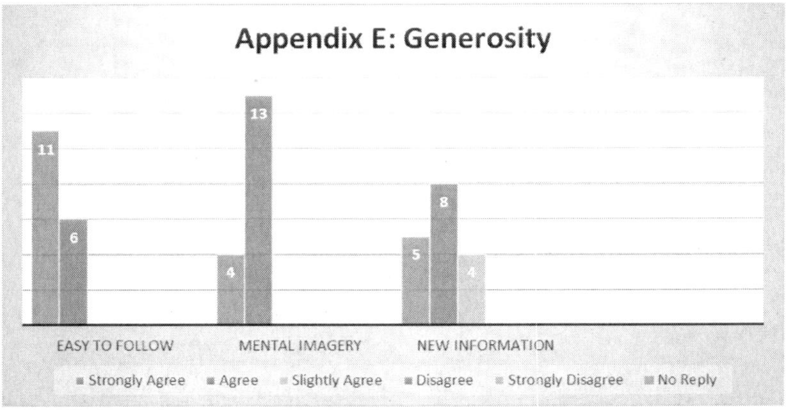

Figure 6.13: Participants' responses to questions on ease in following the class, mental imagery, and new information.

On the need for more sessions on generosity, 1 strongly agreed, 10 agreed, 4 slightly agreed, 1 strongly disagreed, and 1 gave no reply. The session was helpful for participants' reflection on their level of generosity; 7 strongly agreed, 9 agreed, and 1slightly agreed. The session was important for participants' preparation for evangelization; 3 strongly agreed, 13 agreed, and 1 slightly agreed.

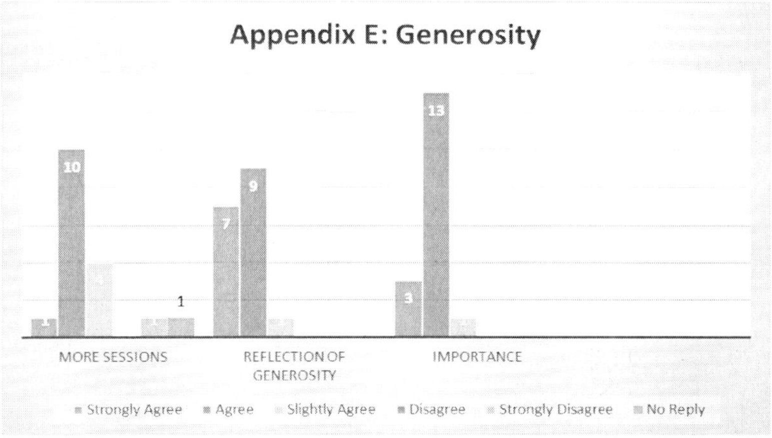

Figure 6.14: Participants' responses to questions on more sessions, reflecting on generosity and its importance.

On the 10-minute rule enhancing participants' approach to generosity, 7 strongly agreed, 6 agreed and 4 slightly agreed. Regarding the importance of generosity in preparation for teamwork, 10 strongly agreed, and 7 agreed. The next question recommended the class on generosity to someone desiring knowledge in generosity; 6 strongly agreed, and 11 agreed.

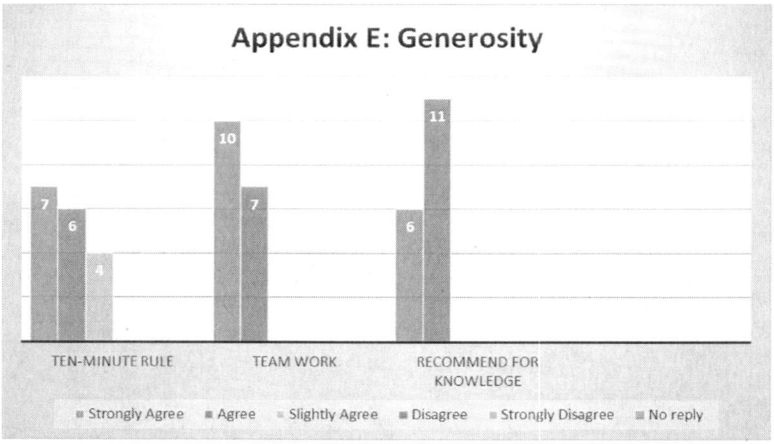

Figure 6.15: Participants' responses to questions on the 10-minute rule, generosity and teamwork, and recommending the class to someone desiring knowledge of generosity.

On using social media in this session to enhance each participant's learning, 4 strongly agreed, 10 agreed and 3 slightly agreed. Regarding recommending the class to someone preparing for evangelization, 9 strongly agreed, and 8 agreed. The overall response on the class on generosity indicates that the participants were engaged and found the class on generosity beneficial for their training; however, while most participants indicated a need for more sessions, one participant strongly disagreed with the need for more sessions.

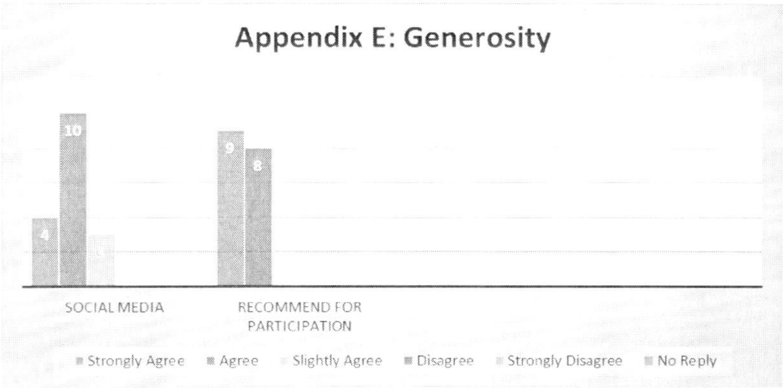

Figure 6.16: Participants' responses to questions on social media use and recommending the class for preparation for evangelization.

Regarding the class on evangelization, participants provided their feedback via a survey. In response to the question, "the session had a clear idea and was easy to follow," 4 strongly agreed that it was easy to follow and had a clear idea, 11 agreed, and 2 gave no reply. The session had examples and mental imagery appealing to a novice in evangelization; 7 strongly agreed, 7 agreed, 1 slightly agreed, and 2 gave no reply. Responding to the session having new information that challenged participants' approach to evangelization, 6 strongly agreed, 9 agreed, and 2 gave no reply.

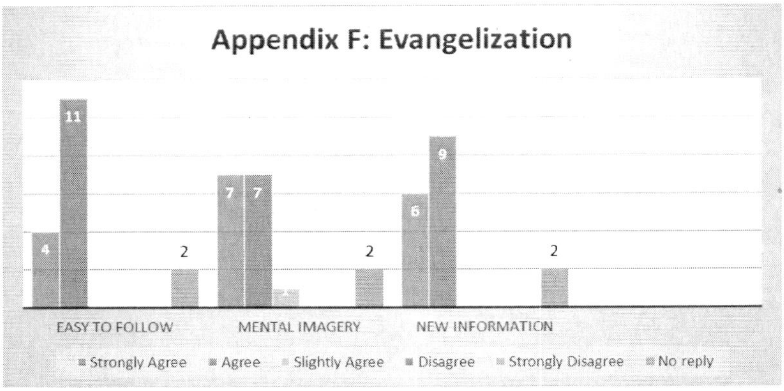

Figure 6.17: Participants' responses to questions on ease in following the class, mental imagery, and new information.

On providing helpful guidelines for evangelization, 6 strongly agreed, 8 agreed, 1 slightly agreed, and 2 gave no reply. For the need for more sessions on evangelization, 5 strongly agreed, 7 agreed, 3 slightly agreed, and 2 gave no reply. For being helpful to distinguish between evangelization and ecumenism, 2 strongly agreed, 7 agreed, 4 slightly agreed, and 4 gave no reply.

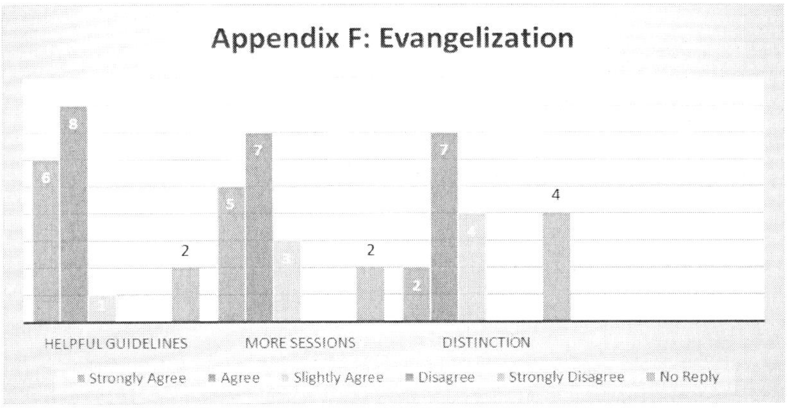

Figure 6.18: Participants' responses to helpful guidelines, more sessions, and distinction.

On the 10-minute rule helping prepare for evangelization, 6 strongly agreed, 9 agreed, and 2 gave no reply. Regarding the session's helpfulness in preparation for teamwork, 4 strongly agreed, 10 agreed 1 slightly agreed, and 2 gave no reply. The next question was on the session being helpful for preparation for evangelization; 4 strongly agreed, 10 agreed, and 3 gave no reply.

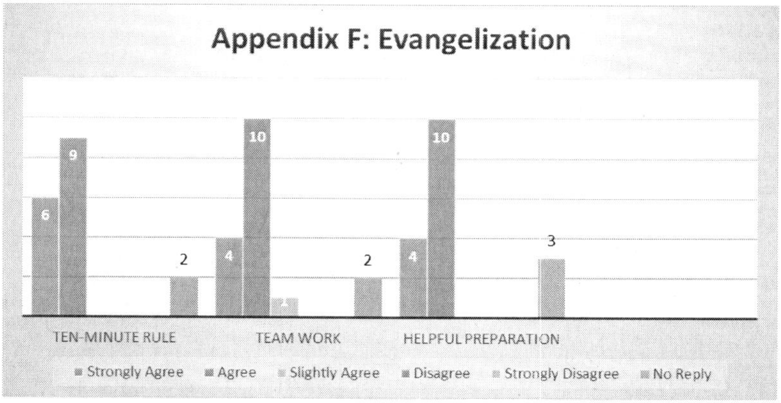

Figure 6.19: Participants' response to questions on the 10-minute rule, evangelization and teamwork, and the class is helpful for evangelization preparation.

On using social media in this session to enhance their learning and participation, 7 strongly agreed, 8 agreed, and 2 gave no reply. Regarding recommending the session to someone desiring to participate in evangelization, 7 strongly agreed, 8 agreed, and 2 gave no reply. The overall response to the class on evangelization indicates that the participants were engaged and found the class beneficial for their training. However, all participants agreed that there was a need for more sessions on evangelization.

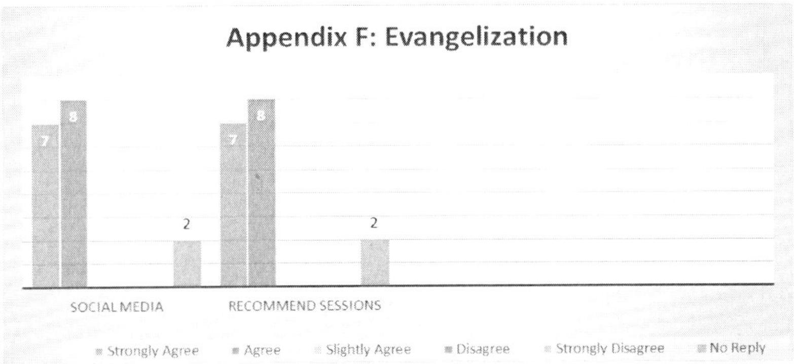

Figure 6.20: Participants' responses to questions on social media use and recommending the class for preparation for evangelization.

In the fifth week of the program, following the online inventory and the discernment of their gifts, I invited participants to reflect on their experience by giving feedback via a survey. In response to the question, the session had clear ideas and was easy to follow; 4 strongly agreed, 10 agreed, 1 slightly agreed, and 2 gave no reply. The session had examples and mental imagery that appeal to a novice in discernment; 5 strongly agreed, 10 agreed, and 2 gave no reply. The session had new information that challenged participants' approach to discernment; 4 strongly agreed, 11 agreed, and 2 gave no reply.

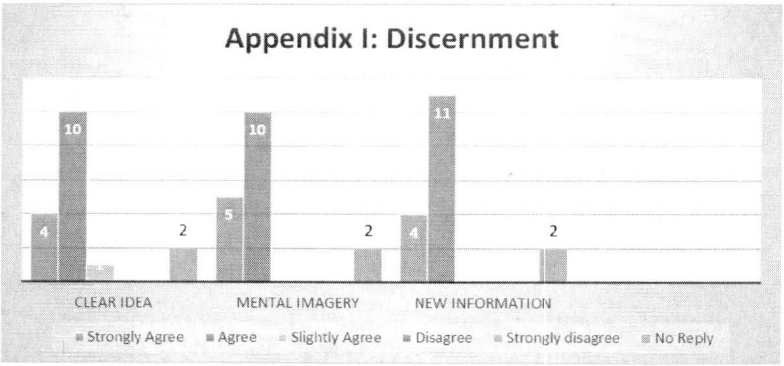

Figure 6.21: Participants' responses to questions on a clear idea, mental imagery, and new information

On the need for more sessions on discernment, 3 strongly agreed, 10 agreed, 2 slightly agreed and 2 gave no reply. On the sessions helping distinguish between a charism and a talent, 6 strongly agreed, 9 agreed, and 2 gave no reply. In response to the discernment of charisms as beneficial for evangelization, 7 strongly agreed, 7 agreed, 1 slightly agreed, and 2 gave no reply.

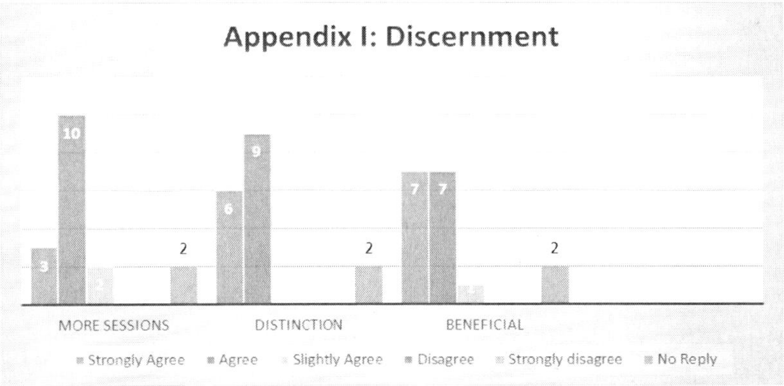

Figure 6.22: Participants' responses to more sessions, distinction, and helpful for evangelization.

In response to the question that identifying one's charisms makes a person more effective to work as part of a team, 6 strongly agreed, 9 agreed, and 2 gave no reply. Regarding the discernment of gifts as good preparation for team building, 6 strongly agreed, 9 agreed, and 2 gave no reply. The next question was whether the social media application enhanced participants' learning and preparation for evangelization; 5 strongly agreed, 10 agreed, and 2 gave no reply. The survey also asked if participants would recommend the discernment session to anyone desiring to participate in evangelization; 5 strongly agreed, 9 agreed 1 slightly agreed, and 2 gave no reply. The overall response on the discernment sessions indicates that the participants were engaged and found the discernment of their charisms beneficial in their training for evangelization.

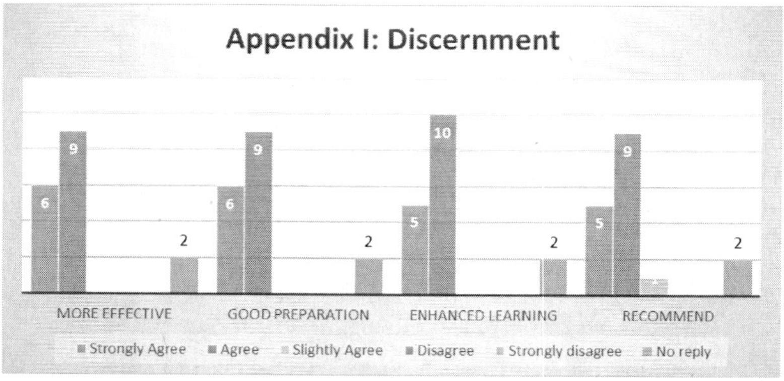

Figure 6.23: Participants' responses on the effectiveness of charisms, helpful in preparing for evangelization, enhancing their learning, and recommending the session to someone who desires to participate in evangelization.

At the end of the nine-week program, participants filled out surveys to assess their learning and the entire program. The goal of this survey was to establish overall learning at the end of the program and discern areas that were beneficial in preparing participants and areas that may need improvement. The charts below reflect participants' responses to the end-of-program survey. In response to the question that the program had clear goals that were easy to follow, 5 strongly agreed, 9 agreed, 1 slightly agreed, and 2 gave no reply. Regarding increased participation in the parish because of the program, 6 strongly agreed, 8 agreed, 1 slightly agreed, and 2 gave no reply. Regarding enhancing their evangelization efforts through social media use like Facebook, 2 strongly agreed, 12 agreed, 2 slightly agreed, and 1 gave no reply.

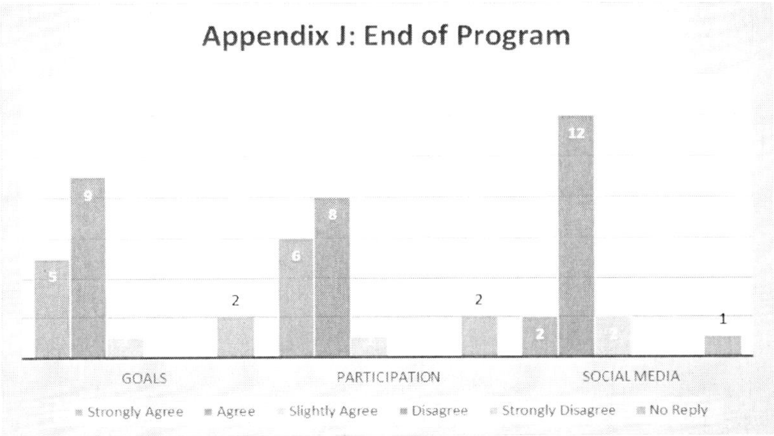

Figure 6.24: Participants' responses on clear goals, participation, and social media use.

Participants also responded that the program's multiple learning formats enhanced their development and growth; 5 strongly agreed, 10 agreed, and 2 gave no reply. The program provided new information that challenged participants' approach to discipleship and evangelization; 6 strongly agreed, 9 agreed, and 2 gave no reply. The program was sufficient for preparing participants for evangelization; 2 strongly agreed, 11 agreed, 2 slightly agreed, and 2 gave no reply.

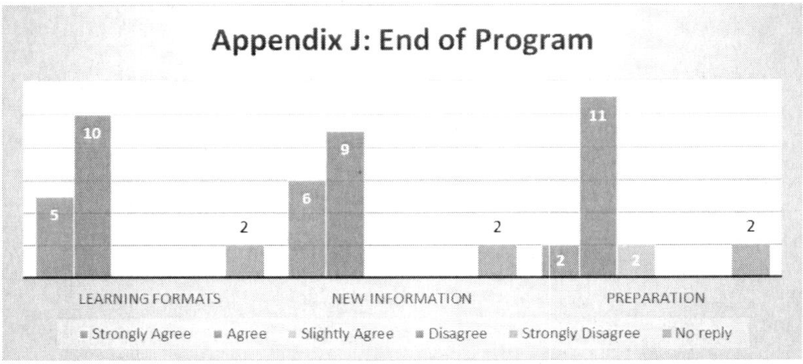

Figure 6.25: Participants' responses on learning formats, new information, and preparation.

Participants responded to increased awareness of their role as lay Catholics due to the program; 6 strongly agreed, 9 agreed, and 2 gave no reply. In response to journaling enhancing participants' growth in the faith, 3 strongly agreed, 10 agreed, 2 slightly agreed, and 2 gave no reply. Regarding participation in the program as helpful in developing in prayer, 4 strongly agreed, 10 agreed, 1 slightly agreed, and 2 gave no reply.

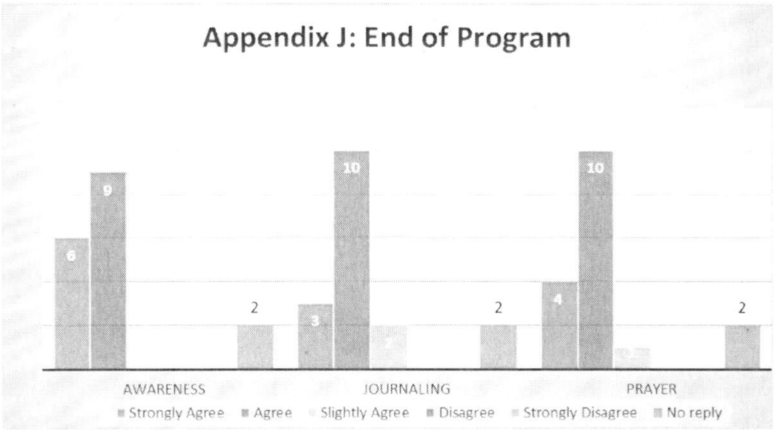

Figure 6.26: Participants' responses to increased awareness, journaling, and prayer.

Participating in the program helped develop a habit of study; 6 strongly agreed, 9 agreed, and 2 gave no reply. The program helped participants grow in generosity; 3 strongly agreed, 10 agreed, 2 slightly agreed, and 2 gave no reply. That scripture reading enhances participation in evangelization; 6 strongly agreed, 6 agreed, 3 slightly agreed, and 2 gave no reply.

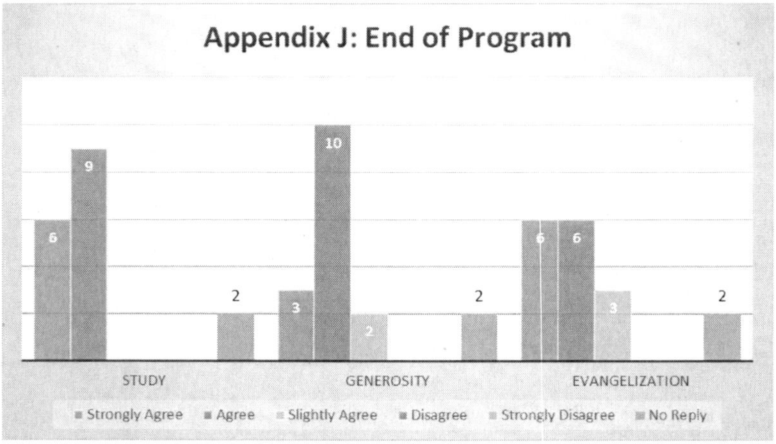

Figure 6.27: Participants' responses to study, generosity and evangelization.

In response to the question that participating in the program helped in the discernment of their charisms, 7 strongly agreed, 8 agreed, and 2 gave no reply. The program was helpful for their engagement in teamwork; 7 strongly agreed, 7 agreed, 1 slightly agreed, and 2 gave no reply. Regarding the social media applications that enhance participants' learning and participation, 4 strongly agreed, 6 agreed, 5 slightly agreed, and 2 gave no reply. Participants were divided on the use of social media to enhance their learning based on the average age of the participants in the program—men and women in their sixties and seventies.

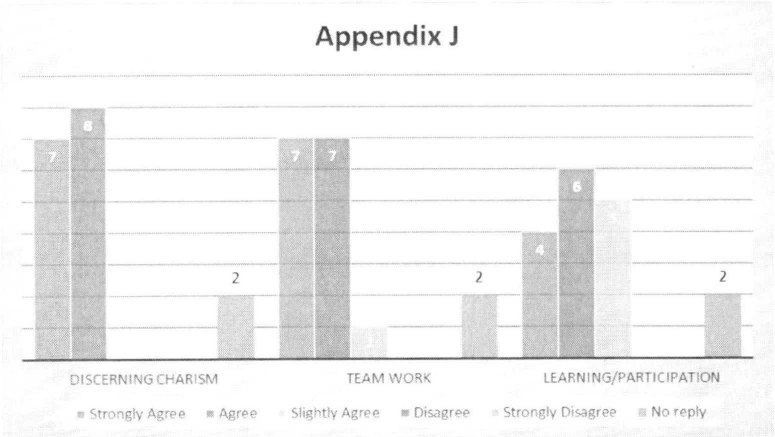

Figure 6.28: Participants' responses to discerning their charisms, teamwork, and their learning/participation.

The participants responded to recommending the sessions to someone desiring to participate in evangelization; 8 strongly agreed, 7 agreed, and 2 gave no reply. In response to the question, "I don't think praying at least ten minutes a day is essential for evangelization," 1 agreed, 8 disagreed, 6 strongly disagreed, and 2 gave no reply. In this question on prayer for at least ten minutes, most participants agreed that it was necessary to prepare them for evangelization. Only one participant did not see it as necessary. Regarding not knowing their charisms before participating in evangelization, 1 agreed, 1 slightly agreed, 7 disagreed, 6 strongly agreed, and 2 gave no reply. Therefore, most participants believed that knowing their charisms before evangelizing was essential.

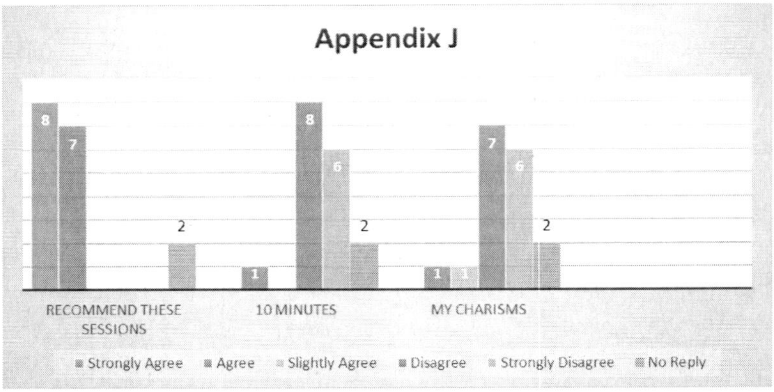

Figure 6.29: Participants' responses to recommending the sessions, ten minutes of prayer, and their charisms.

Responding to the importance of being part of a faith community to evangelize, 9 strongly agreed, 5 agreed, 1 slightly agreed, and 2 gave no reply. And in response to being part of a team facilitating their evangelization participation, 9 strongly agreed, 5 agreed, 1 disagreed, and 2 gave no reply. These questions on being part of a faith community or a team indicate that members were more comfortable evangelizing with a team.

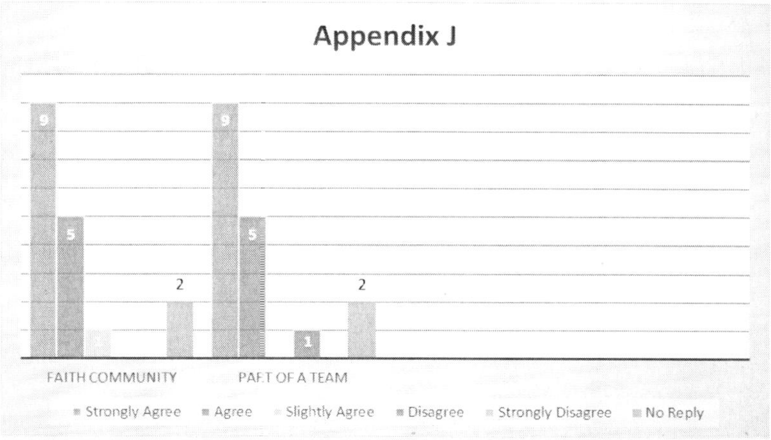

Figure 6.30: Participants' responses to the importance of a faith community and being part of a team.

Participants also had the opportunity to respond to which session or sessions were most engaging in their formation; 9 indicated the session on prayer, 6 indicated the session on study, 5 indicated the session on generosity, 6 indicated the session on evangelization, and 11 indicated the session on the discernment of their charisms. From this assessment, most participants were more engaged in the session on the discernment of their charisms and least engaged in the session on generosity. For me, the researcher, this indicates the challenge that some lay members have in connecting the mental idea of generosity and the practical expression of generosity through the exercise of their gifts. The participants who did not give any reply to the surveys, had emergency engagements that required their transition away from the parish, hence they were unable to return the feedback for the surveys.

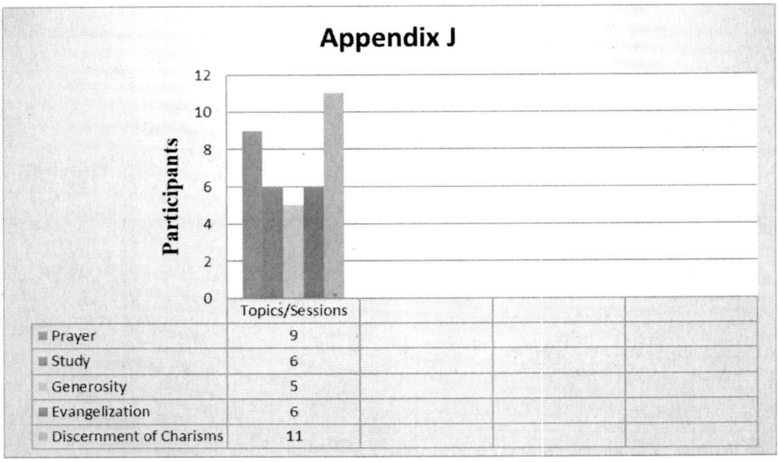

Figure 6.31: Participants' responses to the sessions that were most engaging.

Some participants also shared some additional comments to express their thoughts on the program. One participant noted that "this is a class that every Catholic could benefit from, especially those who are not intimidated sharing the faith." Another participant commenting about the session on the discernment of their charisms noted that "I think it could be of benefit for all parishioners." Another participant commented that "the session on prayer helped my prayer life very much. I think that I will be able to evangelize better." Another participant notes that "I think more thought needs to be given to make this program (in our parish) more successful and satisfying as an online program." Another participant noted that "I liked the interaction between all the members and their charisms." Another participant observed that "I believe if we role-played and were coached before the live recordings, we would be much more comfortable and at ease when being recorded." Another participant added that "I think this course has made us aware that we, as members of a parish, are responsible for going out and bringing people to Christ."

The first six weeks of the program was live streamed to parishioners from Sacred Heart parish. The last three weeks of the program involved a weekly live stream via Facebook to all fifty-six inactive members invited by the group to follow the engagement, including a few parishioners from Sacred Heart. Also, a few members from our neighboring parish, Immaculate Conception, New Madrid, were part of this online audience. From the live stream, I observed that some parishioners, who participated in the online engagement, viewed the events only for a few minutes and did not follow the entire live stream. For the parishioners and inactive members who followed the entire live stream, I invited them to offer their feedback only in the last three weeks. All fifty-six inactive members were invited to offer their feedback, only four returned their feedback via the surveys sent out. I will review the results from the four respondents below.

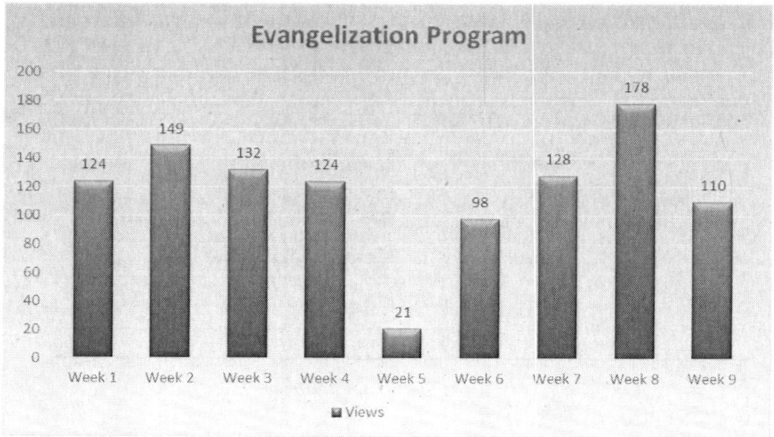

Figure 6.32: Viewership of evangelization and faith sharing engagement program.

Three inactive members and one non-Catholic responded to the surveys sent out to assess their feedback on the evangelization intervention via social media. To the question "if they learned something new about the Catholic faith through the evangelizing sessions; 3 respondents strongly agreed, 1 agreed, and 52 gave no reply. To the question of being comfortable sharing their opinions during the sessions, 3 strongly agreed, 1 agreed, and 52 gave no reply. In response to a follow-up discussion about the Catholic faith, 3 strongly agreed, 1 slightly agreed, and 52 gave no reply.

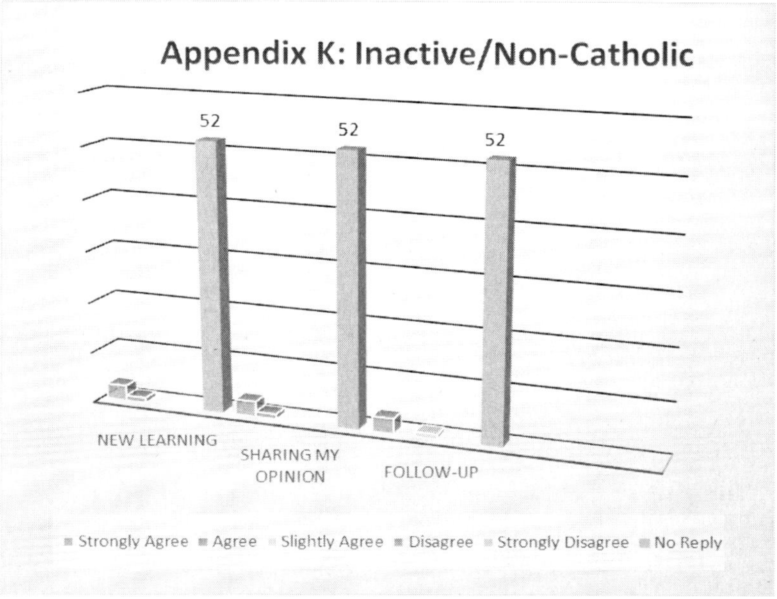

Figure 6.33: Inactive Catholic responses to new learning, Sharing their opinion, and follow-up.

The respondents also gave feedback on the discussions being essential for their transition into the Catholic faith; 3 strongly agreed, 1 disagreed, and 52 gave no reply. One respondent indicated an openness to listen at this time. In response to the enhancement to learning and participation via social media use, all 4 respondents strongly agreed, and 52 gave no reply.

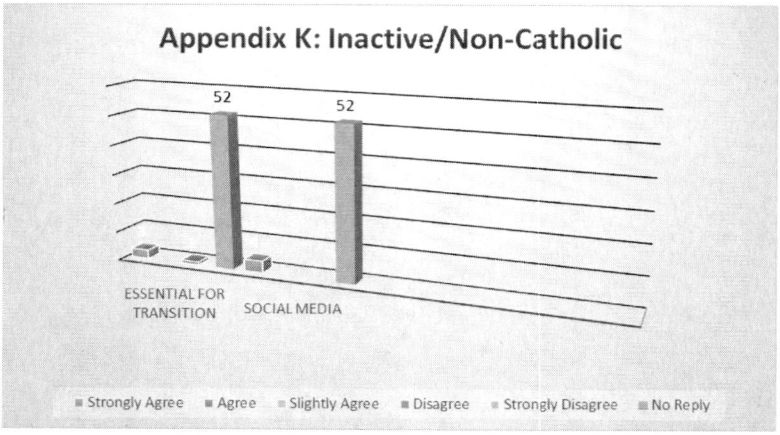

Figure 6.34: Inactive Catholic responses to the faith-sharing session being essential for their transition and social media use.

6.3: General Observation by Participants, Researcher, Followers, and Observers

Two months from the end of the nine-week program, I invited participants for a follow-up survey and interview regarding any changes in their approach to evangelization since their participation in the program. The interview was informal and not in a group format but a face-to-face session with each participant. Participants had some personal comments worth sharing as feedback to evaluate the strength or areas of improvement for the development of the program. The general observation in their words is as follows: one participant noted that "the value of being involved with this group was an education for me." Another participant noted that "I find it easier to profess my faith and to tell others about my love of being Catholic." As regards the charism team, one participant commented that "our group of three worked well together, always open to each other's suggestions." Another participant speaking on engaging the inactive members noted that "one soul back to Christ is worth the work put into getting the gathering arranged." Another participant noted that "we did have four people show up at our gathering (after the evangelization outreach), and that to me is a success. We can continue to encourage members to come back when we encounter them in the community. It is far better to do something than nothing." Another participant noted the following; "I thought the online viewing by others was a great idea, and I hope that, perhaps, we have planted some seeds." Another participant indicated, "I learned not to be afraid of evangelization, but to meet people where they are and with the gifts that I possess, not pushing into something I am not comfortable with (doing). This brings a comfort level in the process that I didn't have before." Another participant added, "I loved the way we all came together with our gifts to try and organize a way we could get others to return to Church. I believe it is a work in progress, and it is going to take more than one meeting to convince some to return. I think we have the tools, and now we must use them." Another participant observed that "I entered this class with great curiosity and enthusiasm. The very fact that the class confirmed my self-knowledge regarding my strengths gave me a degree of confidence I didn't have before." Regarding prayer, one participant noted that "it was also so helpful to learn about the 10-minute rule for prayer. I have made this an important part of my daily spiritual plan."

Moreover, during the informal interview sessions, I invited participants to gauge their experience of the sessions to make a comparative analysis of their present state as lay evangelists. To the question on their previous experience of *lectio divina* and their current engagement of this form of prayer was like; one participant noted, "I always thought of *lectio divina* as reading and studying the bible but not so much as prayer. But now I do." Another participant noted, "I learned about it in reading and research and have tried it over the years—never very successful. This class helped me to understand it better."

Regarding how bible study has strengthened each participant in becoming a better evangelizer, one participant noted, "I think that after participating in the class, I feel more drawn to doing a deep-dive through scripture, as well as reading some of the authors' [sic] recommended in the class." I asked participants how the discernment process affected their journey as Catholics, and one participant responded, "I knew we had gifts and talents given to us, but it was a definite surprise to learn how our particular 'gifts' are to be used." As regards participants' level of confidence at the end of the class, one participant noted, "I feel my charism allows me to build a relationship with people so that they can begin to see the working of God in my life and cause them to be curious about my faith."

When it came to the follow-up surveys, the results revealed the growth that had taken place in the participants' lives in the two months after the formation program enabled them to reflect on the class. The chart below gives a comparative view of their results before and after the class. Responding to their confidence before the class regarding evangelizing inactive Catholics or the unaffiliated and their confidence after the class; 1 participant was confident before the class to evangelize, 15 were confident after the class, and 1 was uncertain. In response to the knowledge of their gifts before the class, 2 were aware of their gifts, and 17 recognized their gifts after the class. As to their familiarity with *lectio divina* before the class and using it as part of their practice of prayer after the class; 9 participants were familiar with *lectio divina* before the class, 10 indicated that they were using it as part of their practice of prayer, and 7 were uncertain of its importance in their life of prayer. Regarding a daily routine of Bible study before and after the class, 10 participants had a daily routine before the class, 12 had a daily routine after the class, 5 participants were uncertain in their response. From these responses, the knowledge of their gifts was beneficial for participants' development, and *lectio divina* does not appeal to almost half of the participants as necessary to their practice of prayer.

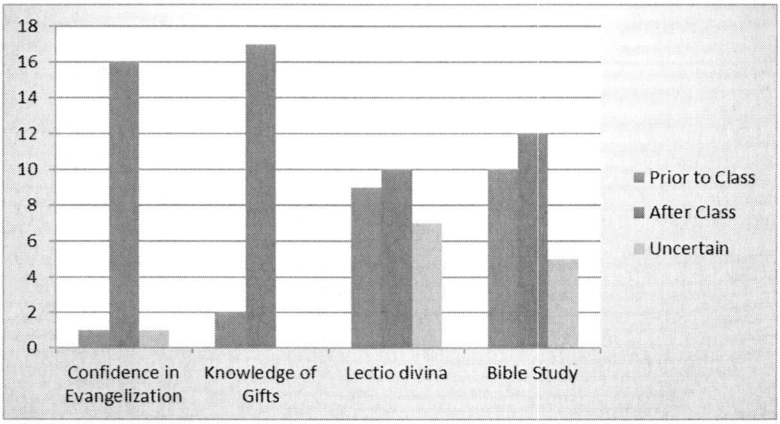

Figure 6.35: Participants' comparative responses to their confidence in evangelizing, knowledge of their gifts, and *lectio divina* and bible study.

At the end of my interviews with the participants, I invited the inactive Catholics for a follow-up after their in-person visit to the parish. In response to my question of one take-away from the faith-sharing sessions, one inactive member shared that "the prayers that each lay participant prayed made a strong impact in their life and the words they shared on the videos were felt. I know that it has challenged me to be more active in participation in my faith." Another inactive member added that "I think it's a good start for inactive members and the unaffiliated because it's not as threatening and people won't be afraid that they are being judged." Another inactive member noted that "I enjoyed these sessions, learning to become a better serving Catholic." The inactive Catholic response shows that they experienced some warmth from the lay evangelists' messages and outreach, and at least two inactive members felt challenged to take the next step of being engaged in their faith.

After the class, I invited members from Immaculate Conception Parish who followed the faith-sharing sessions for their feedback. One member who followed the faith-sharing sessions responded to the question of how the sessions challenged them to approach their faith, noting that "the lay participants' message did help me—mostly to challenge me to active participation in encouraging our fallen-away Catholics to return to the faith. If Sacred Heart parishioners can do this, then I can too!" This response helped summarize the goal of the program of challenging lay Catholics to see their role as evangelizers.

From my notes and observations during the entire process, I observed that some of the participants were uncertain at the start of the program of the goal of this project. The first three weeks seemed somewhat muted in participation, but after the *Called and Gifted* discernment process, participants seemed to come alive with new energy and enthusiasm. Their excitement in recognizing their gifts and working with others as a team turned the tide for them. I was particularly impressed by the bonding within the group—including my increased connection with the participants throughout the process.

6.4: Recommendations and Conclusion

As I look at the entire thesis project, I would like to offer my recommendations based on the original intent of engaging in this project and its future. My thesis set out to address the problem that the parishioners were unaware of their call to evangelize. Simultaneously, some had reservations about the parish's future due to diminishing attendance and not knowing how to address the issue. Therefore, the goal of this project was to address the lack of evangelization and provide training and resources to increase participation and membership within the parish. As I have discussed in my thesis, the process involved developing a training program that facilitated an awareness of evangelization. The program further enhanced intentional discipleship skills within the parish community to engage inactive Catholics and the unaffiliated through social media use.

In the weeks following the training program, I have seen new life within the parish community and the experience of a new kind of excitement among participants. According to many participants, the enthusiasm observed is due to the knowledge of their gifts and skills to interact with the inactive members during the program. Many have suggested that they would appreciate a few sessions inviting the entire parish community to recognize their vocation as disciples and evangelizers. The discernment of gifts is also a significant aspect of growth which participants have suggested will engage members to be more involved within the parish.

As I look towards the next steps for this program, offering regular conferences on prayer, Scripture studies, evangelization, and the discernment of gifts can address some of the deficiencies in this project. These will guide active Catholics to develop skills to evangelize and share their faith. Within a rural parish community, the opportunities may be challenging because of the distance between parishes and negotiating a suitable time. Nevertheless, some sacrifices will be vital to provide frequent opportunities to strengthen active members in their faith. Since my context is in southeast Missouri, I will offer an invitation to parishes within the region to participate in a joint session as a possible way of promoting this program on evangelization.

Moreover, while my hypothesis was that the role of the laity in evangelization is indispensable for the growth and life of the Church, as the Vatican II Documents indicate, this thesis project shows how valid this point is and the need to engage the laity by training them in evangelization. In conclusion, a nine-week program only introduces participants to the mission of evangelization. Still, as members of the laity become proficient in their faith, there is the possibility that they can engage other members of the laity. Also, with the increased engagement that social media offers, I will take advantage of the opportunity to live stream and connect members of the laity who desire to engage the inactive, the unaffiliated, or the nones within the region and beyond. Social media platforms can enhance our evangelization strategy, and I think this is the first step of many to engage our active Catholics, inactive Catholics, and the unaffiliated.

APPENDICES

Appendix A

Bulletin Announcement

Parishioners have discussed the need to increase participation and membership in our parish. To do that, parishioners are invited to participate in a nine-week training program. Kindly take a moment and ask yourself the following questions:

- Are you an active member of the parish interested in learning how to grow in prayer?
- Are you interested in studying the scriptures with other members of the parish?
- Are you interested in knowing your gifts and how you can use it to build up our parish through teamwork and evangelization?
- Are you seeking an opportunity to volunteer in a ministry within the parish or to grow in generosity with your time or talent?
- Do you desire to share your faith with inactive members or the unaffiliated within the community?
- Do you desire to know how you can use social media to evangelize others?

If so, kindly sign up for a nine-week training program on evangelization that will offer you an opportunity to grow in your faith but also enable you to share your faith with either inactive members or the unaffiliated in our community.

Appendix B

Parish Membership Questionnaire

Name:..

Date:...............................

1. Parish Membership. Please check one:
- 1-5 years
- 6-10 years
- 11-15 years
- 20 years or more

2. Depth of involvement in parish life:
- Very involved
- Fairly involved
- Slightly involved
- Involved
- Other.

3. Please Explain your involvement in a few words:

4. What has contributed to your growth within the Church[Please Circle from the list below]:

    a. weekly Adoration
    b. parish community
    c. a parish member
    d. mass and reconciliation
    e. weekly catechesis classes
    f. Personal prayer
    g. Bible Study
    h. other [Please explain in a few words]:

5. Have you had training /formation in any of the following? [Please Circle]

    Prayer/ *Lectio Divina*
    Charism/ Gifts discernments
    Team building
    Evangelization
    Social media, e.g. Facebook, twitter, Instagram, or other media apps.

6. Any other suggestions:

Appendix C

Prayer Response Survey

Please read each statement carefully and respond by circling the response that best describes your reaction to the statement.

Name of session:
Date:
Name of participant:

1. The session had a clear idea that was easy to follow.
   Strongly agree. Agree. Slightly Agree.
   Disagree. Strongly disagree

2. The session had examples and mental imagery that appealed to a novice in prayer.
   Strongly agree. Agree. Slightly Agree.
   Disagree. Strongly disagree

3. The session had new information that challenged my approach to prayer.
   Strongly agree. Agree. Slightly Agree.
   Disagree. Strongly disagree

4. I will like some more sessions on prayer.

   Strongly agree. Agree. Slightly Agree.
   Disagree. Strongly disagree

5. The session helped me make a distinction between private and public prayer.

   Strongly agree. Agree. Slightly Agree.
   Disagree. Strongly disagree

6. Participating at mass is helpful in preparation for evangelization.

   Strongly agree. Agree. Slightly Agree.
   Disagree. Strongly disagree

7. The ten-minute rule has enhanced my prayer life.

   Strongly agree. Agree. Slightly Agree.
   Disagree. Strongly disagree

8. Praying the rosary is essential in my preparation for evangelization.

   Strongly agree. Agree. Slightly Agree.
   Disagree. Strongly disagree

9. I would recommend this session to someone desiring knowledge and growth in prayer.

   Strongly agree. Agree. Slightly Agree.
   Disagree. Strongly disagree

10. The social media application used in this session enhanced my learning/participation.

    Strongly agree. Agree. Slightly Agree.
    Disagree. Strongly disagree

11. I would recommend this session to someone desiring to participate in evangelization.

    Strongly agree. Agree. Slightly Agree.
    Disagree. Strongly disagree

Appendix D

Study Response Survey

Please read each statement carefully and respond by circling the response that best describes your reaction to the statement.

Name of session:

Date:

Name of participant:

1. The session had a clear idea that was easy to follow.

   Strongly agree. Agree. Slightly Agree.

   Disagree. Strongly disagree

2. The session had examples and mental imagery that enhanced my formation.

   Strongly agree. Agree. Slightly Agree.

   Disagree. Strongly disagree

3. The session had new information that challenged my approach to Study.

   Strongly agree. Agree. Slightly Agree.

   Disagree. Strongly disagree

4. I will like some more sessions on Study.

   Strongly agree. Agree. Slightly Agree.
   Disagree. Strongly disagree

5. The session helped me make a distinction between spiritual reading and regular reading.

   Strongly agree. Agree. Slightly Agree.
   Disagree. Strongly disagree

6. Participating in regular study and spiritual reading is helpful in preparation for evangelization.

   Strongly agree. Agree. Slightly Agree.
   Disagree. Strongly disagree

7. The ten-minute rule has enhanced my study of scripture.

   Strongly agree. Agree. Slightly Agree.
   Disagree. Strongly disagree

8. Scripture study is essential for my preparation to evangelize.

   Strongly agree. Agree. Slightly Agree.
   Disagree. Strongly disagree

9. I would recommend this session to someone desiring knowledge in Church teaching.

   Strongly agree. Agree. Slightly Agree.
   Disagree. Strongly disagree

10. The social media application used in this session enhanced my learning/participation.

    Strongly agree. Agree. Slightly Agree.
    Disagree. Strongly disagree

11. I would recommend this session to someone desiring to participate in evangelization.

    Strongly agree. Agree. Slightly Agree.
    Disagree. Strongly disagree

Appendix E

Generosity Response Survey

Please read each statement carefully and respond by circling the response that best describes your reaction to the statement.

Name of session:
Date:
Name of participant:

1. The session had a clear idea that was easy to follow.
   Strongly agree. Agree. Slightly Agree.
   Disagree. Strongly disagree
2. The session had examples and mental imagery that enhanced my formation.
   Strongly agree. Agree. Slightly Agree.
   Disagree. Strongly disagree
3. The session had new information that challenged my approach to generosity.
   Strongly agree. Agree. Slightly Agree.
   Disagree. Strongly disagree
4. I will like some more sessions on generosity.
   Strongly agree. Agree. Slightly Agree.
   Disagree. Strongly disagree

5. The session helped me reflect on my level of generosity.

   Strongly agree. Agree. Slightly Agree.

   Disagree. Strongly disagree

6. The session of generosity is essential in my preparation for evangelization.

   Strongly agree. Agree. Slightly Agree.

   Disagree. Strongly disagree

7. The ten-minute rule has enhanced my approach to generosity.

   Strongly agree. Agree. Slightly Agree.

   Disagree. Strongly disagree

8. The session of generosity is essential in my preparation for teamwork.

   Strongly agree. Agree. Slightly Agree.

   Disagree. Strongly disagree

9. I would recommend this session to someone desiring knowledge of generosity.

   Strongly agree. Agree. Slightly Agree.

   Disagree. Strongly disagree

10. The social media application used in this session enhanced my learning/participation.

    Strongly agree. Agree. Slightly Agree.

    Disagree. Strongly disagree

11. I would recommend this session to someone desiring to participate in evangelization.

    Strongly agree. Agree. Slightly Agree.

    Disagree. Strongly disagree

Appendix F

Evangelization Response Survey

Please read each statement carefully and respond by circling the response that best describes your reaction to the statement.

Name of session:
Date:
Name of participant:

1. The session had a clear idea that was easy to follow.
   Strongly agree. Agree. Slightly Agree.
   Disagree. Strongly disagree
2. The session had examples and mental imagery that appealed to a novice in evangelization.
   Strongly agree. Agree. Slightly Agree.
   Disagree. Strongly disagree
3. The session had new information that challenged my approach to evangelization.
   Strongly agree. Agree. Slightly Agree.
   Disagree. Strongly disagree
4. The session provided helpful guidelines for evangelization.
   Strongly agree. Agree. Slightly Agree.
   Disagree. Strongly disagree

5. I will like some more sessions on evangelization.

   Strongly agree. Agree. Slightly Agree.
   Disagree. Strongly disagree

6. The session helped me make a distinction between evangelization and ecumenism.

   Strongly agree. Agree. Slightly Agree.
   Disagree. Strongly disagree

7. The ten-minute rule has helped my preparation in evangelization.

   Strongly agree. Agree. Slightly Agree.
   Disagree. Strongly disagree

8. Participating in this session was a helpful preparation for teamwork.

   Strongly agree. Agree. Slightly Agree.
   Disagree. Strongly disagree

9. Participating in this session was a helpful preparation for evangelization.

   Strongly agree. Agree. Slightly Agree.
   Disagree. Strongly disagree

10. The social media application used in this session enhanced my learning/participation.

    Strongly agree. Agree. Slightly Agree.
    Disagree. Strongly disagree

11. I would recommend this session to someone desiring to participate in evangelization.

    Strongly agree. Agree. Slightly Agree.
    Disagree. Strongly disagree

Appendix G

The Catholic Spiritual Gifts Inventory[213]

1. When I find myself in a disorganized situation, I help things become better organized.
2. Others have told me that I seem unusually fulfilled and content as a single person.
3. One of the most satisfying ways to express my Christian faith is to create physical objects that are useful or beautiful.
4. I have accurately "recognized" whether a person's words or actions were of divine, demonic, or merely human origin.
5. People tell me that I am a remarkably effective listener or counselor who is able to comfort and heal others by talking with them.
6. Talking about Catholicism with those who have little knowledge of it is one of the most fulfilling and important things I can do.
7. I have trusted God to bring the seemingly impossible to pass and have seen God provide in a remarkable way.
8. It gives me great joy to give as much as I possibly can to those in need.

---

[213] Sherry Weddell, *The Catholic Spiritual Gifts Inventory, Third Edition* (Colorado Springs, CO: The Siena Institute Press, 1998), 11-60.

9. I have seen God heal the sick in a remarkable way through my treatment, care, and/or prayers.
10. It is much more fun and effective for me to work in the background and help someone else succeed than to be "up-front" myself.
11. I know that hospitality is an important part of living as a Christian and I do whatever I can to make guests and visitors feel welcome.
12. Others have told me that my prayers for them have been answered in wonderful or even extraordinary ways.
13. I read and study a great deal in order to understand the universe God has made.
14. One of the most helpful and fulfilling things that I do as a Christian is to find ways in which people can work together to change things for the better.
15. I am drawn to and work well with people who are ignored or rejected by society.
16. People of a different race or culture have told me that they feel very comfortable with me and have been spiritually nourished by my presence and friendship.
17. Singing or playing a musical instrument is one of my favorite ways to pray.
18. I understand the importance of Christian community and do everything I can to make it possible for Christians to spend time together sharing their experiences of following Jesus.

19. I know that it is vital that God's people respond to His call today with their whole hearts; I do whatever I can to make this clear to other Christians.
20. One of the most important ways I express my faith as a Christian is by doing the practical work that it takes to make good things happen.
21. People tell me that I have helped them learn new concepts or skills that made a real difference in their lives.
22. I could live more comfortably but choose not to in order to live like the poor and identify with Jesus.
23. I love the experiential wisdom of the Catholic tradition and enjoy solving practical problems by applying spiritual truths in creative ways.
24. I love to use words creatively and have been told that I am an unusually gifted writer.
25. Providing the planning and coordination that it takes to make good things happen is a very fulfilling way for me to live my faith and serve others.
26. Being single is a positive state of life that enables me to serve God and others more freely.
27. I am particularly good at solving problems related to the creation, repair, or maintenance of material objects.
28. My own faith has been strengthened when I have "sensed" the Spirit of God resting on or dwelling in certain people, places, or things.

29. People talk to me when they are troubled or in need of direction and have told me that they have been greatly comforted or helped as a result.
30. Non-Christians or non-practicing Christians have told me that I have helped them develop a living faith in Christ and his Church.
31. Other people have observed that I am unusually willing to take risks and to step out into the unknown because I have a deep trust in God's love and provision.
32. I give money and things freely to others because God has given so much to me.
33. I have prayed for and/or cared for others and unexpected healing has occurred as a result.
34. Using my gifts to help another person accomplish something significant is a creative experience for me; I have been told that my assistance makes a big difference.
35. People tell me that they have been encouraged and even healed because I have invited them to spend time in my home or with my family and/or friends.
36. I have sensed a "call" to pray intensely for people or places that I didn't know well and have prayed for that situation until I felt that I was "done".
37. I love the rich intellectual heritage of Catholicism, which is an important source of nourishment for my intellectual and spiritual life.

38. Others have told me that I am a person of unusual vision and direction and have worked with me to make a vision happen.
39. I identify with the pain of those who suffer or are oppressed and seek to comfort them with the love of Jesus through my actions.
40. I move freely outside the culture of my childhood in order to share Christ with more people.
41. I am musically gifted; my ability to write or perform music has delighted and nourished others and given me much satisfaction.
42. People have told me that they have been spiritually encouraged and strengthened by being part of a small Christian group that I facilitated.
43. I have known what God wanted to say to certain people in particular situations and have been told that others were encouraged or moved to action because I shared the message I had been given.
44. I'd rather be active and do something practical for God or other people than just sit around talking.
45. I enjoy spending time learning new things and then figuring out how to help others learn them as well.
46. I am readily accepted by the poor because I live simply.
47. I make good practical use of information or ideas that I learn and have helped others do the same in their own lives.

48. Writing is a good way to pray and is one of my favorite spiritual exercises.
49. I have been told that my ability to organize has enabled others to better accomplish a complex task that served God and people.
50. I readily identify with the celibate lifestyle of many of the saints.
51. I enjoy making beautiful or useful things, especially when I know that it benefits others and gives glory to God.
52. I have sensed that a person, place, or thing was affected by an evil spirit and have had this sense confirmed by later events or other evidence.
53. Individuals who talk to me about their lives tell me that they grow personally and spiritually as a result of our time together.
54. I have shared how God's presence and love has blessed my life in a way that is meaningful or intriguing to non-Christians or non-practicing Christians.
55. I am very confident that God answers prayer and will provide all that I need to serve others and fulfill my God-given mission in life.
56. I give sacrificially when confronted with the urgent needs of other people, the Church, or groups that work for worthy causes.

57. Others have told me that God has used me to heal them of serious physical or mental illness when the chances for recovery were small.
58. Other people have told me that my personal assistance enabled them to achieve things that they could never have done by themselves.
59. gladly open my home and give food and shelter to others, even when I am not prepared to have guests.
60. Praying for others is one of my favorite spiritual practices; seeing God answer my prayers in such wonderful ways has greatly strengthened my faith.
61. I enjoy discovering "new" ideas or truths for myself by reading or observing situations first-hand.
62. I know that providing the sense of direction and hope that enables a group of people to accomplish new things is one of the most effective ways in which I serve God and others.
63. People who are struggling with suffering or injustice have told me that I have been a real source of practical help and comfort.
64. I am frustrated when I have little opportunity to relate to or work among people of a different language or cultural background.
65. Writing or performing music is a wonderful creative experience for me; others have told me that they really enjoy my music.

66. One of the best ways that I can serve God and others is by doing whatever it takes to enable a group of Catholics to come together and explore their faith.
67. When I share truths from Scripture or Catholic teaching with others, I do it in a way that calls for a personal response and sparks worship, repentance, or action in individuals or groups.
68. I am able to see unmet organizational needs or "holes" that others miss and enjoy doing what it takes to fill the "hole" and make a difference.
69. People tell me that I am able to explain complicated matters in a way that is clear, easy to remember, and interesting.
70. I am drawn to the spiritual challenge of living simply in a materialistic culture.
71. Other people tell me that I am very perceptive and talk their problems or decisions over with me in order to take advantage of my insights.
72. I find writing to be a very satisfying and creative means of expressing my faith.
73. I enjoy and am unusually good at coordinating the many "pieces" of a larger task or goal.
74. I am glad that I am single because I am freer to serve God.
75. Others have noticed that I am unusually competent and creative in dealing with my physical surroundings.

76. Others acknowledge that I accurately "recognized" the presence of God's Spirit or an evil spirit in a person, place, or thing in the absence of tangible evidence.
77. One of the most satisfying ways to express my faith is by listening and talking to those who seek out my encouragement, nurture, or counsel.
78. I enjoy sharing with non-Catholics the joy and peace which my Catholic faith has given me.
79. When making important life decisions, I rely upon the guidance, love, and providence of God to an unusual degree.
80. I do without things myself in order to be able to give more to others.
81. I know that many of the sick that I care and/or pray for will be healed in remarkable ways.
82. Personally providing the practical assistance that enables someone else to achieve good things is one of the best ways that I can serve God and help people.
83. Strangers and newcomers have told me that I have made them feel especially welcome and comfortable.
84. God seems to use my prayers as a way through which his love can reach others; people are often helped in remarkable ways when I pray for them.
85. Others have told me that the insights that I have gained from my study have helped them grow in their own intellectual and spiritual lives.

86. I am able to communicate my ideas or vision for a better future in a way that motivates others to work to help make the vision happen.
87. One of the most important ways I express my Catholic faith is through practical deeds of love and care for the poor, the sick, or those suffering from injustice.
88. I have learned a new language well in order to be an effective agent of God's purposes among people of a different culture than my own.
89. I often express my faith in and love for God through music.
90. I enjoy and have been told that I am particularly good at guiding and encouraging the spiritual growth of a group of Christians.
91. I identify with the ministry and experience of the Old Testament prophets or of prophetic saints because God seems to use me in the same way.
92. Other people tell me that I really know what it takes to get a job done, and that my personal willingness to do whatever is necessary makes a big difference.
93. I have explained aspects of the Catholic faith to others and have seen them "get it" and grow spiritually as a result.
94. Others have told me I have sacrificed a great deal materially in order to follow Jesus and be close to the poor.
95. I have helped others make good decisions that reflected the purposes of God in the midst of uncertainty.

96. People really enjoy my writing and have told me that their lives have been touched or enriched by reading things that I have written.
97. An important way I express my faith is by providing the planning and coordination that enables good things to be accomplished.
98. I am emotionally and spiritually fulfilled without the special intimacy of marriage.
99. Working creatively with my hands is a form of prayer or contemplation for me.
100. I have been able to help others because I could tell the difference between unusual human behavior caused by emotional disturbance and that influenced by a demonic presence.
101. Others have said that talking to me has really helped them grow spiritually and/or discern God's direction for their life.
102. I have shared my faith with people who have no church background and have seen them desire to know more about the Catholic faith as a result.
103. When I or others close to me are in need, I am excited by the opportunity to trust in God and see how wonderfully he will provide.
104. I am so confident that God will meet my own financial needs that I am free to give sacrificially.

105. I have been used by God to restore strength and health to someone who suffered from a serious emotional or physical illness.
106. One of my favorite ways to live out my faith is to use my abilities to personally help someone else accomplish a task or lifework to which God has called him or her.
107. I express my Catholic faith by welcoming guests or strangers to my home or community as I would welcome Jesus.
108. I know from experience that praying is one of the most helpful and satisfying things that I can do for someone else.
109. Study is an important spiritual exercise for me because when I study anything, I know that I am contemplating an aspect of God's loving activity in the universe.
110. I have successfully led a group through the process of turning a faith-based vision or goal into reality.
111. I find it very satisfying to personally serve those who are sick, powerless, or rejected by society and have been told that my efforts make a real, practical difference in their lives.
112. I enjoy worshipping and sharing my faith with Catholics of another ethnic or cultural background.
113. Others have told me that my music has given them a lot of pleasure and/or drawn them closer to God.

114. I have helped a specific group of Christians grow spiritually by bringing people together on a regular basis to pray, study, or share their faith.
115. Others have told me that I have really helped them by sharing an encouraging or challenging word from God with them at a critical moment in their life.
116. I am a jack-of-all-trades, willing to do whatever it takes to serve God and others and to make good things happen.
117. One of my favorite ways to express my Catholic faith is by helping others learn truths, information, or skills that will help them grow spiritually and as human beings.
118. I identify with the saints who embraced poverty for Christ's sake.
119. I often come up with good, faith-based solutions to complex problems.
120. One of the best ways I can express my human experience and/or Christian faith is through writing.

Appendix H

Charisms

- Administration
- Celibacy
- Craftsmanship
- Discernment of Spirits
- Encouragement
- Evangelism
- Faith
- Giving
- Healing
- Helps
- Hospitality
- Intercessory Prayer
- Knowledge
- Leadership
- Mercy
- Missionary
- Music
- Pastoring
- Prophecy
- Service
- Teaching
- Voluntary Poverty
- Wisdom
- Writing

Appendix I

Discernment Survey

Please read each statement carefully and respond by circling the response that best describes your reaction to the statement.

Name of session:
Date:
Name of participant:

1. The session had clear ideas that were easy to follow.
   Strongly agree. Agree. Slightly Agree.
   Disagree. Strongly disagree
2. The session had examples and mental imagery that appealed to a novice in discernment.
   Strongly agree. Agree. Slightly Agree.
   Disagree. Strongly disagree
3. The session had new information that challenged my approach to discernment.
   Strongly agree. Agree. Slightly Agree.
   Disagree. Strongly disagree
4. I will like some more sessions on discernment.
   Strongly agree. Agree. Slightly Agree.
   Disagree. Strongly disagree

5. The session helped me make a distinction between charisms and talents.

   Strongly agree. Agree. Slightly Agree.
   Disagree. Strongly disagree

6. Participating in the discernment of charisms is beneficial for evangelization.

   Strongly agree. Agree. Slightly Agree.
   Disagree. Strongly disagree

7. Identifying my charism or charisms made me more effective to work as part of a team

   Strongly agree. Agree. Slightly Agree.
   Disagree. Strongly disagree

8. The session is good preparation for team building.

   Strongly agree. Agree. Slightly Agree.
   Disagree. Strongly disagree

9. The social media application used in this session enhanced my learning/participation.

   Strongly agree. Agree. Slightly Agree.
   Disagree. Strongly disagree

10. I would recommend this session to someone desiring to participate in evangelization.

    Strongly agree. Agree. Slightly Agree.
    Disagree. Strongly disagree

Appendix J

End of Program Survey

Please read each statement carefully and respond by circling the response that best describes your reaction to the statement.

Name of session:
Date:
Name of participant:

1. The program had clear goals that were easy to follow.
   Strongly agree. Agree. Slightly Agree.
   Disagree. Strongly disagree
2. The program increased participation in the parish.
   Strongly agree. Agree. Slightly Agree.
   Disagree. Strongly disagree
3. The use of social media like Facebook enhanced our evangelization efforts.
   Strongly agree. Agree. Slightly Agree.
   Disagree. Strongly disagree
4. The program had multiple learning formats that enhanced my development and growth.
   Strongly agree. Agree. Slightly Agree.
   Disagree. Strongly disagree

5. The program had new information that challenged my approach to discipleship and evangelization.

    Strongly agree. Agree. Slightly Agree.
    Disagree. Strongly disagree

6. The program was sufficient for my preparation for evangelization.

    Strongly agree. Agree. Slightly Agree.
    Disagree. Strongly disagree

7. The program increased the awareness of my role as a lay Catholic.

    Strongly agree. Agree. Slightly Agree.
    Disagree. Strongly disagree

8. Journaling enhances my growth in the faith.

    Strongly agree. Agree. Slightly Agree.
    Disagree. Strongly disagree

9. Participating in the program was helpful in developing in prayer.

    Strongly agree. Agree. Slightly Agree.
    Disagree. Strongly disagree

10. Participating in the program was helpful in developing in study.

    Strongly agree. Agree. Slightly Agree.
    Disagree. Strongly disagree

11. Participating in the program was helpful in growing in generosity.

    Strongly agree. Agree. Slightly Agree.
    Disagree. Strongly disagree

12. Participating in Scripture reading enhances my participation in evangelization.

    Strongly agree. Agree. Slightly Agree.
    Disagree. Strongly disagree

13. Participating in the program was helpful in discerning my charism.

    Strongly agree. Agree. Slightly Agree.
    Disagree. Strongly disagree

14. Participating in the program was helpful in engaging in teamwork.

    Strongly agree. Agree. Slightly Agree.
    Disagree. Strongly disagree

15. The social media applications used in the sessions enhanced my learning/participation.

    Strongly agree. Agree. Slightly Agree.
    Disagree. Strongly disagree

16. I would recommend these sessions to someone desiring to participate in evangelization.

    Strongly agree. Agree. Slightly Agree.
    Disagree. Strongly disagree

17. I don't think praying at least 10 minutes a day is essential for evangelization.

    Strongly agree. Agree. Slightly Agree.
    Disagree. Strongly disagree

18. I don't need to know my charisms to participate in evangelization.

    Strongly agree. Agree. Slightly Agree.
    Disagree. Strongly disagree

19. Being part of a faith community is essential for evangelization.

    Strongly agree. Agree. Slightly Agree.
    Disagree. Strongly disagree

20. Being part of a team facilitates my participation in evangelization.

    Strongly agree. Agree. Slightly Agree.
    Disagree. Strongly disagree

21. What topics/ sessions were most engaging to you?
    - Prayer
    - Study
    - Generosity
    - Evangelization
    - Discernment of Charisms

22. Kindly add any other comments that you feel will benefit the program in the future:

Appendix K

Survey for Inactive/Non-Catholic Respondents

Name:

Date:

1. I learned something new about the Catholic Faith in the faith-sharing sessions. Please check one.
   - Strongly agree
   - Agree
   - Slightly Agree
   - Disagree
   - Strongly disagree
2. I was comfortable sharing my opinion in the sessions. Please check one.
   - Strongly agree
   - Agree
   - Slightly Agree
   - Disagree
   - Strongly disagree
3. I want a follow-up discussion about the Catholic Faith. Please check one.
   - Strongly agree
   - Agree
   - Slightly Agree
   - Disagree

- Strongly disagree

4. The discussions I had were essential for my transition back/ into the Catholic faith. Please <u>check one</u>.
   - Strongly agree
   - Agree
   - Slightly Agree
   - Disagree
   - Strongly disagree
5. The social media applications used in each session enhanced my participation/learning

   - Strongly agree
   - Agree
   - Slightly Agree
   - Disagree
   - Strongly disagree
6. Do you have any suggestions on topics/areas you would like to learn?

Appendix L

Informed Consent Form for Participants

Introduction: My name is Fr Dominic Ibok; I am a Catholic priest and a student at the Aquinas Institute conducting a research study in fulfillment of a Doctorate in Ministry. My telephone number is 314-489-4220. My research Supervisor is Sr. Sara Fairbanks, O.P. and her phone number is 314-256-8800. You may contact either of us at any time if you have questions about this study.

Purpose: This research aims to facilitate a training program for evangelization at Sacred Heart Church for active members.

Procedure: If you consent, you will be asked several questions in an oral interview that will take place either over the phone or in person.

Time required: The Interview will take approximately 30-45 minutes

Voluntary Participation: Your participation in this study is completely voluntary. If you choose to participate, you may still refuse to answer any question that you do not wish to answer. You may also withdraw from the study at any time.

Risks: There are no known risks associated with this interview. However, you might feel distressed in the course of the conversation. If this happens, please inform me promptly.

Benefits: While there is no guaranteed benefit, it is possible that you will enjoy sharing your answers to these questions or that you will find the conversation meaningful. This study is intended to benefit the parish by enlivening our discourse on the theology and practice of evangelization.

Confidentiality/Anonymity: Your name will be kept confidential in all reporting and/or writing related to this study. I will be the only person present for the interview, and when I write the research results, I will use pseudonyms – made-up names – for all participants, unless you specify in writing that you wish to be identified by name.

If you wish to choose your own pseudonym for the study, please indicate the first name you would like me to use for you here: ______________________________.

Sharing the results: I plan to construct an ethnography – a written account of what I learn – based on these interviews together with my reading and research. This will be submitted to my research supervisor. I also plan to share what I learn from the study with the participants and the parish. Portions of the research may be printed and made available to the participants.

Publication: There is the possibility that I will publish this study or refer to it in published writing in the future. In this event, I will continue to use pseudonyms (as described above), and I may alter some identifying details to protect your anonymity further.

Before you sign: By signing below, you agree to an interview for this research study. Be sure that any questions you may have are answered to your satisfaction. If you agree to participate in this study, a copy of this document will be given to you.

Participant's signature:

Date:

Print Name:

Researcher's signature:

Date:

Print Name:

Appendix M

Interview Questions for Participants

1. What knowledge of *lectio divina* did you have before the class, and are you using *lectio divina* after participating in the class?
2. Has your prayer life before the class involved at least 10 minutes of prayer a day, and are you practicing the 10-minute rule after participating in the class?
3. What kind of daily Bible study routine did you have before the class, and do you have a daily routine of bible study after participation in the class?
4. What awareness of your gifts or charisms did you have before the class, and did you realize any new gifts you possess due to the class?
5. How confident were you in evangelization before the class? And do you feel greater confidence now to evangelize inactive members after you participated in the class?
6. What skills did you have for evangelization before the class, and did you learn any new skills to facilitate your effort in evangelization after the class?
7. How did you view social media use as a platform for evangelization before the class? And do you think social media is an essential platform for evangelization after the class?

8. Kindly share one significant take-away after participating in the class.

Appendix N

Follow-up Interview Questions for Inactive Catholics/Observers

1. How did the lay participants' message assist your growth in faith and active participation? Kindly explain in a few words.

2. Kindly share why you may or may not be open to being engaged in the future.

3. Kindly identify one significant take-away from the faith-sharing encounter.

Bibliography

Aljentera, Clarissa Valbuena. *The Parish Guide To Social Media: How Social Networking Can Recharge Your Ministry.* New London: Twenty-third publications, 2013.

Barber, Terry. *How to Share Your Faith With Anyone: A Practical Manual for Catholic Evangelization.* San Francisco: Ignatius Press, 2013.

Barron, Robert. *Catholicism: A Journey to the Heart of the Church*. New York, NY: Image Publishing, 2011.

________. *Catholicism: The New Evangelization.* Des Plaines: Word on Fire Publications, 2013.

________. *Exploring Catholic Theology: Essays on God, Liturgy, and Evangelization.* Grand Rapids, MI: Baker Academic, 2015.

________. *Arguing Religion: A Bishop Speaks at Facebook and Google.* Park Ridge, IL: Word on Fire, 2018.

Barron, Robert, and John Allen, Jr. *To Light a Fire on the Earth: Proclaiming the Gospel in a Secular Age.* New York, NY: Image Publishing, 2017.

Benedict XVI. *Ubicumque Et Semper.* Apostolic Letter Establishing the Pontifical Council for Promoting the New Evangelization. September 21 2010. Available online at http://www.vatican.va/content/benedict-xvi/en/apost_letters/documents/hf_ben-xvi_apl_20100921_ubicumque-et-semper.html.

________. *Address to Participants in the Plenary Assembly of the Pontifical Council for Promoting the New Evangelization.* May 30 2011. Available online at http://www.vatican.va/content/benedict-xvi/en/speeches/2011/may/documents/hf_ben-xvi_spe_20110530_nuova-evangelizzazione.html.

________. *The New Evangelization: Building the Civilization of Love.* December 12 2000. Available online at https://www.piercedhearts.org/benedict_xvi/Cardinal%20Ratzinger/new_evangelization.htm.

________. *The Priest and Pastoral Ministry in a Digital World: New Media at the Service of the Word.* May 16 2010. Available online at http://w2.vatican.va/content/benedict-xvi/en/messages/communications/documents/hf_ben-xvi_mes_20100124_44th-world-communications-day.html.

________. *Silence and Word: Path of Evangelization.* 20 May 2012. Available online at http://w2.vatican.va/content/benedict-xvi/en/messages/communications/documents/hf_ben-xvi_mes_20120124_46th-world-communications-day.html.

Boguslawski, Steven and Ralph Martin, Editors. *The New Evangelization: Overcoming the Obstacles.* Mahwah: Paulist Press, 2008.

Boyack Kenneth, CSP, editor. *Catholic Evangelization Today: A New Pentecost for the United States.* New York: Paulist Press, 1987.

Brown, Raymond E., S.S. The Churches the Apostles Left Behind. Mahwah, NY: Paulist Press, 1984.

Catechism of the Catholic Church: Revised in Accordance with the Official Latin Text Promulgated by Pope John Paul II. 2nd ed. Vatican City: Libreria Editrice Vaticana, 1997.

Cavadini, John C. and Donald Wallenfang, eds. *Pope Francis and the Event of Encounter: Global Perspectives on the New Evangelization.* Eugene, OR: Pickwick Publications, 2018.

Chautard, Jean-Baptiste, O.C.S.O. *The Soul of the Apostolate.* Charlotte: Tan Books, 1946.

Collins, Raymond F.. The Many Faces of the Church: A Study in New Testament Ecclesiology. New York, NY: Crossroad Publishing Company, 2003.

Committee on Evangelization and Catechesis, United States Conference of Catholic Bishops. *Disciples Called to Witness the New Evangelization.* Washington D.C.: United States Conference of Catholic Bishops, 2013.

Connors, Michael E., ed. *We Preach Christ Crucified.* Collegeville, MN: Liturgical Press, 2014.

Dawson, Steve. *Catholic Street Evangelization: Stories of Conversion and Witness.* San Francisco: Ignatius Press, 2016.

Dulles, Avery Cardinal. Models of the Church. New York, NY: Image Books, 2002.

________. *Evangelization for the Third Millennium.* Mahwah, NY: Paulist Press, 2009.

Earley, David, and Rod Dempsey. *Disciple Making is...: How to Live the Great Commission with Passion and Confidence.* Nashville, TN: B&H Publishing Group, 2013.

Fowler, James W. *Stages of Faith: The Psychology of Human Development and the Quest for Meaning.* San Francisco, CA: Harper and Row Publishing, 1981.

Friedman, Thomas. *The World is Flat: A Brief History of the Twenty-first Century.* New York, NY: Farrar, Straus and Giroux, 2005.

Gaillardetz, Richard R. *Rediscovering Vatican II: The Church in the Making.* Mahwah, NJ: Paulist Press, 2006.

________. *The Cambridge Companion To Vatican II.* Boston, MA: Cambridge University Press, 2020.

Gaillardetz, Richard R. and Catherine E. Clifford. *Keys to The Council: Unlocking the Teaching of Vatican II.* Collegeville, MN: Liturgical Press, 2012.

Gan, Eugene. Infinite Bandwidth: Encountering Christ in the Media. Steubenville, OH: Emmaus Road Publishing, 2010.

Gittins, Anthony J. *The Way of Discipleship: Women, Men, and Today's Call to Mission.* Collegeville, MN: Liturgical Press, 2016.

Gould, Meredith. *Transcending Generations: A Field Guide To Collaboration in Church.*, Collegeville: Liturgical Press, 2017.

________. *The Social Media Gospel: Sharing the Good News in New Ways.* 2nd Edition. Liturgical Press, 2015.

________. *The Word Made Fresh: Communicating Church and Faith Today.* Harrisburg: Morehouse Publishing, 2008.

Gray, Mark M., and Mary L. Gauthier. *Catholic New Media Use in the United States, 2012.* Center for Applied Research in the Apostolate. Available online at http://www.usccb.org/about/communications/upload/Catholic_New_Media_Use_in_United_States_2012.pdf

Hahn, Scott. *Evangelizing Catholics: A Mission Manual for the New Evangelization.* Huntington, IN: Our Sunday Visitor Publishing, 2014.

Hardon John, S.J. *The Urgent Call to Evangelization.* Accessed online on December 1, 2019. http://www.therealpresence.org/archives/Media/Media_003.htm.

Harrington, Daniel, J, SJ. *Meeting St. cchew Today: Understanding the Man, His Mission, and His Message.* Chicago, IL: Loyola Press, 2010.

________. The Church According to the New Testament: What the Wisdom and Witness of Early Christianity Teach Us Today. Franklin, WI: Sheed and Ward, 2001.

Harrington, Jay M. *Matthew 28: 16-20 and the Feast of the Ascension: An Investigation into a Biblico-Liturgical Relationship*. Chicago, IL: Catholic Theological Union, 1982.

Heil, John Paul. *The Death and Resurrection of Jesus: A Narrative-Critical Reading of Matthew 26-28.* Eugene, OR: Wipf and Stock, 2004.

Hilkert, Mary Catherine, O.P. *Naming Grace: Preaching and the Sacramental Imagination.* New York, N.Y.: Continuum Publishing, 1997.

Hinnebusch, Paul, OP. *St. Matthew's Earthquake: Judgment and Discipleship in the Gospel of Matthew.* Ann Arbor, MI: Servant Books, 1980.

Ivereigh, Austen. *How to Defend the Faith Without Raising Your Voice: Civil Responses to Catholic Hot-Button Issues.* Huntington: Our Sunday Visitor Inc., 2012.

John Paul II. *Catechesi Tradendae.* Apostolic Exhortation on Catechesis in Our Time. October 16 1979. Available online at http://www.vatican.va/content/john-paul-ii/en/apost_exhortations/documents/hf_jp-ii_exh_16101979_catechesi-tradendae.html.

________. *Redemptoris Missio.* Encyclical on the Permanent Validity of the Church's Missionary Mandate. December 7 1990. Available online at http://www.vatican.va/content/john-paul-ii/en/encyclicals/documents/hf_jp-ii_enc_07121990_redemptoris-missio.html.

________. *Proclaiming Christ in the Media at the Dawn of the New Millennium.* June 4 2000. Available online at http://w2.vatican.va/content/john-paul-ii/en/messages/communications/documents/hf_jp-ii_mes_20000124_world-communications-day.html.

________. *Novo Millenio Ineunte.* Apostolic Letter at the Close of the Great Jubilee of the Year 2000. January 6 2001. Available online http://www.vatican.va/content/john-paul-ii/en/apost_letters/2001/documents/hf_jp-ii_apl_20010106_novo-millennio-ineunte.html.

________. *Internet: A New Forum for Proclaiming the Gospel.* May 12 2002. Available online at http://w2.vatican.va/content/john-paul-ii/en/messages/communications/documents/hf_jp-ii_mes_20020122_world-communications-day.html.

Kaczor, Christopher. *The Seven Big Myths About the Catholic Church: Distinguishing Facts from Fiction in Catholicism.* San Francisco: Ignatius Press, 2012.

Kelly, Matthew. *Rediscover Catholicism: A Spiritual Guide To Living With Passion And Purpose.* Hebron, KY: Beacon Publishing, 2010.

________. *The Seven Pillars of Catholic Spirituality.* Hebron, KY: Beacon Publishing, 2011.

________. *The Four Signs of A Dynamic Catholic.* Hebron, KY: Beacon Publishing, 2012.

________. *Perfectly Yourself: Discovering God's Dream for You*. North Palm Beach, FL: Blue Sparrow Publishing, 2017.

________. *Dig the Well before you get Thirsty*. North Palm Beach, FL: Blue Sparrow Publishing, 2020.

Kemp, Carrie and Donald Pologruto. *Catholics Coming Home: A Journey of Reconciliation*. New York, NY: HarperCollins Publishers, 1990.

Laurence, Trevor. *3 Dangers of Preaching in a Digital Age*. The Gospel Coalition. July 29, 2016. Accessed online on May 1, 2019. https://www.thegospelcoalition.org/article/3-dangers-of-preaching-in-a-digital-age/

Leslie, Jennifer. *55 Social Media Marketing Stats For 2018*. Last modified February 1, 2018. Accessed August 6, 2019. https://keap.com/business-success-blog/marketing/social-media/best-social-media-marketing-stats-and-facts

Mallon, James. *Divine Renovation: Bringing Your Parish from Maintenance to Mission*. New London, CT: Twenty-Third Publications, 2014.

Martin, Curtis. *Made for More*. North Palm Beach, FL: Wellspring Publishing, 2008.

________. *Making Missionary Disciples: How to Live the Method Modeled by the Master*. North Palm Beach, FL: Wellspring Publishing, 2018.

Martin, Ralph. *The Urgency of the New Evangelization: Answering the Call.* Huntington, IN: Our Sunday Visitor Publishing, 2013.

Martin, Ralph and Peter Williamson, eds. *John Paul II and the New Evangelization: How You Can Bring the Good News to Others.* Cincinnati: Franciscan Media, 2006.

McAfee, Shaun. *Social Media Gospel: A No-Nonsense Guide To The Proper Use Of Social Media.* St. Louis: En Route Books And Media, 2018.

McLuhan, Marshall and Quentin Fiore. *The Medium is the Message: An Inventory of Effects.* Corte Madera, CA: Wired Books, 1995.

Monshau, Michael, O.P. *The Responsibility of Every Christian: To Preach Jesus Christ.* In *Listening: Journal of Religion and Culture.* The Catholic Recovery of Preaching. Vol. 38, No. 1, Winter 2003, 21-33. Edited by Mark McVann, F.S.C. Ann Arbor, MI: Lewis University, 2002.

Moschella, Mary Clark. *Ethnography as a Pastoral Practice: An Introduction.* Cleveland: The Pilgrim Press, 2008.

Paul VI. Apostolic Exhortation on Evangelization in the Modern World. *Evangelii nuntiandii* (8 December 1975). Frederick, MD: The Word Among Us Press, 1975.

________. *Social Communications and Evangelization in Today's World.* May 16 1974. Available online at http://w2.vatican.va/content/paul-vi/en/messages/communications/documents/hf_p-vi_mes_19740516_viii-com-day.html.

Peterson, Tom. *Catholics Come Home: God's Extraordinary Plan for Your Life*. New York: Image Publishing, 2013.

Pius XII. Encyclical Letter on the Advancement of Motion Pictures, Radio, and Television. *Miranda porsus*. Vatican Website. Accessed online May 1, 2019 at http://w2.vatican.va/content/pius-xii/en/encyclicals/documents/hf_p-xii_enc_08091957_miranda-prorsus.html.

Pontifical Council for Social Communications. Pastoral Instruction *Aetatis Novae*. (February 22, 1992).Vatican Website. Accessed online on May 1, 2019. http://www.vatican.va/roman_curia/pontifical_councils/pccs/documents/rc_pc_pccs_doc_22021992_aetatis_en.html.

Pope Francis. *Lumen Fidei*. Encyclical Letter on Faith. June 29 2013. Available online at http://www.vatican.va/content/francesco/en/encyclicals/documents/papa-francesco_20130629_enciclica-lumen-fidei.html.

________. *Evangelii Gaudium*. Apostolic Exhortation on the Proclamation of the Gospel in Today's World. November 24 2013. Available online at http://www.vatican.va/content/francesco/en/apost_exhortations/documents/papa-francesco_esortazione-ap_20131124_evangelii-gaudium.html.

________. *The Joy of Discipleship: Reflections from Pope Francis on Walking with Christ.* James P. Campbell, ed. Chicago, IL: Loyola Press, 2016.

________. *Communication and Mercy: A Fruitful Encounter.* January 24, 2016. Available online at http://w2.vatican.va/content/francesco/en/messages/communications/documents/papa-francesco_20160124_messaggio-comunicazioni-sociali.html.

Radcliffe, Timothy, OP. *Why Go To Church: The Drama of the Eucharist.* New York: Continuum Books, 2008.

Reid, Barbara E. *The Gospel According to Matthew: Part Two.* Matthew 17-28. Collegeville, MN: Liturgical Press, 2019.

Rivers, Robert S., CSP. From *Maintenance to Mission: Evangelization and the Revitalization of the Parish.* Mahwah, NJ: Paulist Press, 2005.

Schroeder, Roger P. *What is the Mission of the Church: A Guide for Catholics.* Mary Knoll, NY: Orbis Books, 2018.

Senior, Donald, CP. *The Gospel of Matthew.* Nashville, TN: Abingdon Press, 1997.

Senior, Donald, CP and Carroll Stuhlmueller, CP. *The Biblical Foundations for Mission.* Maryknoll, NY: Orbis Books, 1983.

Sri, Edward, and Curtis Martin. *The Real Story: Understanding the Big Picture of the Bible.* North Palm Beach, FL: Wellspring Publishing, 2012.

Synod of Bishops XIII Ordinary General Assembly. *The New Evangelization for the Transmission of the Christian Faith.* June 19 2012. Available online at http://www.vatican.va/roman_curia/synod/documents/rc_synod_doc_20120619_instrumentum-xiii_en.html.

Thompson, William G. *Matthew's Story: Good News for Uncertain Times.* Mahwah, NJ: Paulist Press, 1989.

Troeger, Thomas H. *Ten Strategies For Preaching In A Multimedia Culture.* Nashville: Abingdon Press, 1996.

United States Conference of Catholic Bishops. *A Strategic Plan: Encountering the Mercy of Christ and Accompanying His People with Joy.* Accessed online on December 8, 2019. http://www.usccb.org/about/upload/Strategic-Plan-2017-2020-abbreviated.pdf.

_________. *A Four Year Journey with Christ: Faith, Worship, Witness.* Accessed online on December 8, 2019. http://www.usccb.org/about/2013-2016-plan-suggested-parish-diocesan-roadmap.cfm.

_________. *Living as Missionary Disciples.* Accessed online on December 8, 2019. http://www.usccb.org/beliefs-and-teachings/how-we-teach/catechesis/catechetical-sunday/living-disciples/upload/7-571-missionary-disciples-ndc-worksheets.pdf.

________. *Called and Gifted for the Third Millennium.* Washington D.C.: United States Catholic Conference Inc., 1995.

________. *Go and Make Disciples: A National Plan and Strategy for Catholic Evangelization in the United States.* Washington D.C.: United States Conference of Catholic Bishops, Inc., 2002.

________. *Co-Workers in the Vineyard of the Lord: A Resource for Guiding the Development of Lay Ecclesial Ministry.* Washington D.C.: United States Conference of Catholic Bishops, 2008.

Vatican Council II. Decree on Social Media, *Inter mirifica* (4 December 1963). In *Vatican Council II Constitutions and Declarations: A Completely Revised Translation in Inclusive Language.* Edited by Austin Flannery. Northport NY: Costello Publishing Co., 1996.

________. Dogmatic Constitution on the Church. *Lumen gentium* (21 November 1964). In *Vatican Council II Constitutions and Declarations: A Completely Revised Translation in Inclusive Language.* Edited by Austin Flannery. Northport NY: Costello Publishing Co., 1996.

________. Decree on the Apostolate of the Laity. *Apostolicam actuositatem* (18 November 1965). In *Vatican Council II Constitutions and Declarations: A Completely Revised Translation in Inclusive Language.* Edited by Austin Flannery. Northport NY: Costello Publishing Co., 1996.

________. Dogmatic Constitution *Dei verbum* (18 November 1965). In *Vatican Council II Constitutions and Declarations: A Completely Revised Translation in Inclusive Language*. Edited by Austin Flannery. Northport NY: Costello Publishing Co., 1996.

________. Pastoral Constitution on the Church in the Modern World. *Gaudium et spes* (7 December 1965). In *Vatican Council II Constitutions and Declarations: A Completely Revised Translation in Inclusive Language*. Edited by Austin Flannery. Northport NY: Costello Publishing Co., 1996.

________. Decree on the Mission Activity of the Church. *Ad gentes* (7 December 1965). In *Vatican Council II Constitutions and Declarations: A Completely Revised Translation in Inclusive Language*. Edited by Austin Flannery. Northport NY: Costello Publishing Co., 1996.

Vogt, Brandon. *The Church and New Media: Blogging Converts, Online Activists, and Bishops Who Tweet*. Huntington: Our Sunday Visitor, 2011.

Weddell, Sherry. *The Catholic Spiritual Gifts Inventory*. Colorado Springs, CO: The Siena Institute Press, 1998.

________. *Discerning Charisms: A Workbook for Navigating the Discernment Process*. Colorado Springs, CO: The Siena Institute Press, 2002.

________. *Forming Intentional Disciples: The Path to Knowing and Following Jesus*. Huntington, IN: Our Sunday Visitor Publishing, 2012.

________. *Becoming A Parish of Intentional Disciples.* Huntington, IN: Our Sunday Visitor Publishing, 2015.

________. *Fruitful Discipleship: Living the Mission of Jesus in the Church and the World.* Huntington, IN: Our Sunday Visitor Publishing, 2017.

________. *Called and Gifted Discernment Process: Complete Resource.* Colorado Springs, CO: The Siena Institute Press, 2019.

Wright, Stephen. *Preaching with the Grain of Scripture.* Biblical Grove Series. Cambridge, MA: Grove Books, 2001.

BIOGRAPHY OF THE AUTHOR

NAME	Dominic O. Ibok
BORN:	7 April 1976, Lagos, Nigeria
EDUCATION:	Immaculate Conception High School Seminary Mfamosing, Calabar, Cross State, Nigeria 1986–92 Lagos Baptist Academy, Obanikoro, Lagos, Nigeria 1992–93
	St. Joseph Major Seminary, Ikot Ekpene, affiliate of Federal University of Uyo, Akwa Ibom, Nigeria and Pontifical Urban University, Rome, Italy (B.A., Philosophy/ B. Philosophy) 1995–99
	Kenrick School of Theology, St. Louis, MO (M.A., M.Div.) 2003–07
	Aquinas Institute of Theology, St. Louis, MO (D.Min.) 2016–21

RELIGIOUS PROFESSION AND ORDINATION:

Final Profession, Society of Our Mother of Peace
Ordination as Deacon
Ordination as Presbyter

MINISTRY: Coordinator of Evangelization
St. Elizabeth, Mother of John the Baptist Parish, St. Louis, MO

Religious Superior/ Student Master
Society of Our Mother of Peace, Nigeria

Associate Pastor
Giles Parish, Oak Park, IL
St. Joseph Parish, Springfield, MO

Pastor
Immaculate Conception Parish, New Madrid, MO
Sacred Heart Parish, Caruthersville, MO
St Eustachius Parish, Portageville, MO
St. Cecilia Parish, Kennett, MO

Chaplain

Southeast Correctional Center, Charleston, MO 2019

Missouri Army National Guard, MO 2020 –

PRESENT CHURCH RELATIONSHIP:

Roman Catholic Diocese of Springfield – Cape Girardeau, Missouri

Made in United States
Cleveland, OH
05 November 2026

47477871R00167